AF664477

CHROMATOGRAPHY AND CENTRIFUGATION METHODS

A Beginner's Handbook

THE AUTHOR

Dr. Timir Tripathi is a Professor of Molecular Biology in the Department of Zoology at North-Eastern Hill University (NEHU), Shillong. Previously, he served as the Regional Director of Indira Gandhi National Open University (IGNOU), Kohima. Prior to that role, he held the position of Senior Assistant Professor and Principal Investigator at the Department of Biochemistry, NEHU, Shillong. He holds a Ph.D. from Jawaharlal Nehru University, New Delhi, while conducting research at the Central Drug Research Institute, Lucknow. He has held visiting faculty positions at ICGEB, New Delhi, and Khon Kaen University, Thailand. With over 18 years of experience in teaching and research at the postgraduate and doctoral levels, he has made significant contributions to the understanding of protein structure, function, and dynamics. He has also developed and improved methods for investigating and analyzing proteins. As a principal investigator, he has successfully handled 12 research projects funded by national and international agencies. Dr. Tripathi has been honoured with numerous national and international awards and is a respected member of the National Academy of Sciences, India, and the Royal Society of Biology, UK. He has published over 100 research papers, reviews, and editorial articles in international journals. Additionally, he has edited five books and authored two textbooks for undergraduate and postgraduate students.

CHROMATOGRAPHY AND CENTRIFUGATION METHODS

A Beginner's Handbook

Timir Tripathi

Professor, Department of Zoology,
North-Eastern Hill University, Shillong - 793022, India

2024

Daya Publishing House®

A Division of

Astral International Pvt. Ltd.

New Delhi – 110 002

© 2024 AUTHOR
ISBN: 978-93-5461-731-7

Publisher's Note:

Every possible effort has been made to ensure that the information contained in this book is accurate at the time of going to press, and the publisher and author cannot accept responsibility for any errors or omissions, however caused. No responsibility for loss or damage occasioned to any person acting, or refraining from action, as a result of the material in this publication can be accepted by the editor, the publisher or the author. The Publisher is not associated with any product or vendor mentioned in the book. The contents of this work are intended to further general scientific research, understanding and discussion only. Readers should consult with a specialist where appropriate.

Every effort has been made to trace the owners of copyright material used in this book, if any. The author and the publisher will be grateful for any omission brought to their notice for acknowledgment in the future editions of the book.

All Rights reserved under International Copyright Conventions. No part of this publication may be reproduced, stored in a retrieval system, or transmitted in any form or by any means, electronic, mechanical, photocopying, recording or otherwise without the prior written consent of the publisher and the copyright owner.

Published by : **Daya Publishing House®**
A Division of
Astral International Pvt. Ltd.
– ISO 9001:2015 Certified Company –
4736/23, Ansari Road, Darya Ganj
New Delhi-110 002
Ph. 011-43549197, 8130496929
E-mail: info@astralint.com
Website: www.astralint.com

Digitally Printed at : **Replika Press Pvt. Ltd.**

FOREWORD

It is with great pleasure that I introduce you to "Chromatography and Centrifugation Methods: A Beginner's Handbook." In today's fast-paced world of scientific discovery and technological advancements, the need for a book on efficient separation methods is more critical than ever before. This handbook aims to be a beacon of knowledge, guiding both novices and seasoned professionals through the fascinating realms of chromatography and centrifugation.

Separation methods play a vital role in a wide range of scientific disciplines and industries. The ability to separate and analyse complex mixtures is fundamental to advancing research, quality control, and product development. This book provides an extensive overview of various separation techniques, with a specific emphasis on chromatography and centrifugation, which are the cornerstones of modern separation science.

The beauty of this book lies in its accessibility. Dr. Tripathi has carefully crafted the content to cater to beginners, ensuring that complex concepts are explained in a clear and concise manner. The book takes a step-by-step approach, gradually introducing the fundamental principles, techniques, and applications of chromatography and centrifugation. By following along, readers will gain a solid foundation and a comprehensive understanding of these powerful separation methods.

The breadth and depth of topics covered in this book are impressive. From the basic principles and types to the intricacies of sample preparation and instrument selection, every aspect is thoroughly explored. Furthermore, Dr. Tripathi has delved into the applications of these techniques across various fields, including biotechnology, clinical diagnostics, and industrial sectors. This comprehensive coverage makes the book an invaluable resource for anyone involved in separation science.

I commend Dr. Tripathi for his expertise, diligence, and dedication in compiling this book. His passion for the subject matter shines through, and their commitment to bridging the gap between theory and practice is evident in every chapter. His efforts have resulted in a book that not only imparts knowledge but also inspires curiosity and a desire to explore the limitless possibilities of chromatography and centrifugation.

I extend my gratitude to Dr. Tripathi for this contribution. I am confident that this book will find a rightful place on the bookshelves of students, researchers, and faculty alike. May this book serve as a trusted guide, empowering readers to unravel the mysteries of separation methods and unlock new avenues of scientific discovery.

Prof. Vikash Kumar Dubey
Dean R&D
Indian Institute of Technology (BHU), Varanasi

PREFACE

Separation methods, particularly chromatography and centrifugation, are fundamental to numerous scientific disciplines, including biochemistry, chemistry, biotechnology, life sciences, pharmaceuticals, environmental science, and more. The ability to separate and analyse complex mixtures is crucial for research, quality control, and various industrial processes.

This book, "Chromatography and Centrifugation Methods: A Beginner's Handbook," provides comprehensive information on various separation methods, particularly focussing on chromatography and centrifugation. It aims to be a guide in understanding and harnessing the power of chromatography and centrifugation. The book is designed to be accessible to beginners and students while offering valuable insights for researchers and faculty in the above mentioned fields.

The book is divided into 13 chapters, each covering a specific aspect of chromatography and centrifugation methods. Chapter 1 provides an introduction to separation methods. Chapter 2 discusses the historical background, basic principles of chromatography, and the different types of chromatographic techniques. Chapters 3 to 10 discuss various chromatographic methods, including paper, thin layer, gel filtration, affinity, hydrophobic interaction, gas chromatography, and high-pressure liquid chromatography, respectively. Chapter 11 covers the other chromatographic methods. Chapter 12 focuses on various centrifugation methods. Chapter 13 provides an overview of other separation methods. In the end, a glossary is provided with the common terms used in chromatography and centrifugation.

This book will be a valuable resource for any student interested in chromatography and centrifugation methods, whether a beginner or an experienced researcher. It provides a comprehensive overview of the basic principles, methodology, and applications of various techniques. Throughout the book, clear explanations and illustrative diagrams have been provided to enhance readers' understanding of the subject matter. The latest advancements are also discussed, ensuring that the readers are up to date with the current trends in the field.

I hope the book will serve as a valuable resource in readers' journey of exploring the fascinating world of chromatography and centrifugation. Whether the reader is a student, researcher, or industry professional, I believe this book will empower them with the knowledge and skills necessary to tackle complex separation challenges and contribute to advancements in various scientific fields. I hope that the readers find this book informative, engaging, and inspiring as they embark on their exploration of the two fundamental and widely used separation methods.

Prof. Timir Tripathi

NEHU, Shillong

CONTENTS

01 Introduction to Separation Methods

CHAPTER SUMMARY

- Separation methods refer to diverse techniques used to separate components from mixtures. They play a crucial role in various fields, enabling the purification, analysis, and isolation of desired substances.
- These methods include chromatography, centrifugation, filtration, distillation, extraction, evaporation, crystallization, precipitation, electrolysis, and solvent extraction.
- Factors affecting separation techniques involve the physical and chemical properties of the sample, such as size, density, solubility, reactivity, polarity, pH, as well as parameters like temperature, pressure, flow rate, and column dimensions.
- Separation methods have vast applications in biotechnology, pharmaceuticals, environmental analysis, etc. They are crucial for sample purification, analysis, and separation of components for various industries and scientific research.
- A technique is a specific procedure used to perform a task, while a method is a systematic approach that incorporates multiple techniques to achieve an objective or solve a problem.

1. Introduction

- To analyse different components of a mixture, their separation is necessary.
- To separate a component from a mixture, various separation methods are used. The separation methods are based on the physical and chemical properties of the molecule.
- Separation methods are employed to isolate and purify individual components from complex mixtures.
- These methods aim to separate substances based on their distinct physical or chemical properties, allowing for the extraction of specific components for further analysis, purification, or industrial applications.
- By utilizing various principles and methods, separation methods facilitate the isolation of substances, enabling scientists to better understand their characteristics and use them effectively.

- ☆ The separation methods are used in both scientific research and industrial sectors.

  In research, separation methods play a crucial role in studying the properties, behaviour, and interactions of different substances. They enable the identification and quantification of components within a mixture, allowing researchers to unravel complex chemical and biological processes. Moreover, separation methods serve as the foundation for numerous analytical methods, enabling scientists to perform accurate and reliable measurements.

- ☆ In the industrial sector, separation methods are essential for the production and purification of a wide range of products. Industries such as pharmaceuticals, food and beverage, petroleum, and environmental engineering rely on separation methods to extract, isolate, and refine specific substances. These methods help ensure product quality, remove impurities, and comply with regulatory standards.
- ☆ From ancient civilizations' rudimentary filtration to the alchemists' discovery of distillation, the field has witnessed continuous advancements. The emergence of chromatography in the 19th century revolutionized analysis, while the industrial revolution drove the need for efficient separation processes. The 20th century brought advances in instrumentation and the discovery of new separation principles, expanding applications in various fields. In recent years, the focus has shifted towards sustainable and environmentally friendly methods.
- ☆ The separation methods have evolved, providing invaluable tools for scientific research, quality control, and industrial processes.

2. Types of Separation Methods

- ☆ There are several separation methods available. Each technique has its advantages, limitations, and specific applications.
- ☆ The choice of the appropriate separation technique depends on the nature of the mixture and the desired outcome.
- ☆ Understanding these different methods enables scientists and engineers to effectively separate and purify substances for various scientific, industrial, and research purposes.

2.1. Chromatography

- ☆ Chromatography is one of the most important separation methods. It is based on the differential interactions of components with a stationary phase and a mobile phase.

- ☆ It is widely used for qualitative and quantitative analysis, purification, and separation of complex mixtures.
- ☆ Different types of chromatography employ different stationary and mobile phases to achieve separation.

2.2. Centrifugation

- ☆ Centrifugation is a technique that separates components of a mixture based on their different densities or sedimentation rates.
- ☆ The mixture is placed in a centrifuge, where it is subjected to high-speed rotation. This generates centrifugal force, causing denser components to move towards the bottom while lighter components remain at the top.
- ☆ Centrifugation is used in various fields, including biology, chemistry, and medicine, for the separation and purification of solids, liquids, and cells.

2.3. Filtration

- ☆ Filtration is a technique used to separate solid particles from a fluid or gas by passing the mixture through a porous medium.
- ☆ The porous medium, such as filter paper or a membrane, allows the fluid or gas to pass through while retaining the solid particles.
- ☆ Filtration is widely used in water treatment, air purification, and laboratory applications.

2.4. Distillation

- ☆ Distillation is a technique that separates components of a mixture based on their different boiling points.
- ☆ The mixture is heated, and the vapour is condensed and collected, resulting in the separation of components with different boiling points.
- ☆ Distillation is commonly used in the production of alcoholic beverages, petroleum refining, and laboratory purification processes.

2.5. Extraction

- ☆ Extraction involves the separation of a substance from a mixture using a solvent.
- ☆ The solvent selectively dissolves the desired component, allowing for its separation from the mixture.
- ☆ This technique is widely used in chemistry, pharmaceuticals, and natural product extraction processes.

2.6. Evaporation

☆ Evaporation is a separation technique that exploits the differences in volatility or vapour pressure of components in a mixture.

☆ The mixture is heated, and the component with a lower boiling point evaporates, leaving behind the more volatile component.

☆ Evaporation is commonly used in the production of salt, concentration of solutions, and recovery of solvents.

2.7. Crystallization

☆ Crystallization is a technique that separates a solute from a solvent by inducing the formation of solid crystals.

☆ The solute is dissolved in a solvent, and as the solvent is evaporated or cooled, the solute molecules come together to form crystals.

☆ Crystallization is used for the purification and isolation of substances in industries such as pharmaceuticals and chemical manufacturing.

2.8. Precipitation

☆ Precipitation is a technique that involves the formation of a solid precipitate from a solution by introducing a precipitating agent.

☆ The precipitating agent reacts with the dissolved substances, causing them to form insoluble solids.

☆ Precipitation is used in wastewater treatment, mineral processing, and analytical chemistry.

2.9. Electrolysis

☆ Electrolysis utilizes an electric current to drive a chemical reaction.

☆ It involves the use of electrodes and an electrolyte solution to facilitate the decomposition of compounds into their constituent ions, allowing for their selective separation and purification.

2.10. Solvent extraction

☆ Solvent extraction utilizes the difference in solubility of substances in immiscible solvents.

☆ It involves the extraction of a desired compound from a mixture by selectively dissolving it in a suitable solvent, enabling its separation and purification.

3. Factors Affecting Separation

☆ The success and efficiency of separation methods depend on several factors that influence the separation process.

- Understanding and considering these factors is essential for selecting and optimizing separation methods in scientific and industrial applications.
- By assessing the physical and chemical properties of the sample and considering the factors specific to the chosen technique, scientists and engineers can achieve efficient and reliable separations, leading to accurate and meaningful results.
- The factors that affect separation can be categorized into physical properties of the sample, chemical properties of the sample, and operational factors.

3.1. Physical properties of the sample

- **Size:** The size of the particles or molecules in the sample plays a significant role in separation methods. Methods like filtration and centrifugation are particularly influenced by particle size, where smaller particles may pass through filters or settle more slowly in a centrifuge.
- **Density:** The density of the sample components affects methods such as centrifugation and sedimentation. Components with different densities will separate based on their buoyancy or sedimentation rates.
- **Solubility:** The solubility of substances in different solvents determines the choice of the separation technique. Methods like extraction and chromatography rely on differences in solubility to achieve separation.
- **Boiling point:** Distillation is a separation technique that utilizes the differences in boiling points of components in a mixture. The separation is based on the vaporization and condensation of substances with different boiling points.
- **Melting point:** Crystallization is influenced by the melting points of substances. Components with different melting points can be separated through controlled cooling and crystallization processes.

3.2. Chemical properties of the sample

- **Reactivity:** The reactivity of sample components determines the choice of the separation technique. Components that undergo chemical reactions during separation may require specific methods, such as chemical derivatization or selective precipitation.
- **Polarity:** The polarity of substances affects separation methods such as chromatography. Polar and non-polar components can be separated based on their differential interactions with the stationary phase.
- **pH:** The pH of the sample can influence separation methods, especially in ion exchange chromatography and precipitation reactions. The pH can affect the ionization state and solubility of components.

3.3. Operational factors

- **Temperature:** Temperature plays a crucial role in separation methods such as distillation and extraction. Controlling the temperature allows for the separation of components with different boiling points or solubilities.
- **Pressure:** Pressure affects methods like distillation and supercritical fluid extraction. It can influence the boiling points of substances and the solubility of components in supercritical fluids.
- **Flow rate:** In chromatography, the flow rate of the mobile phase influences the separation process. An optimal flow rate ensures proper interaction between the sample components and the stationary phase.
- **Column length:** The length of the separation column in methods like chromatography affects the separation efficiency. A longer column provides more time for separation but may result in longer analysis times.
- **Column diameter:** The diameter of the separation column also impacts the separation efficiency. A smaller diameter column provides higher resolution but may have limitations in sample loading capacity.
- **Mobile phase composition:** In methods like chromatography, the composition of the mobile phase affects the separation. The choice of solvents and additives can optimize the selectivity and resolution of the separation.

4. Applications of Separation Methods

- Separation methods are vital tools in various scientific disciplines and industries, enabling the isolation and purification of substances from complex mixtures.
- These methods find wide-ranging applications in fields such as pharmaceuticals, environmental analysis, food processing, forensic sciences, and many more.
- The ability to separate and purify substances has profound implications for research, development, and quality control processes.
- Below are a few applications of separation methods. Detailed applications of various methods are discussed in individual chapters.

4.1. Drug discovery and development

- Separation methods are integral to drug discovery and development processes.

- ☆ High-throughput screening methods employ chromatographic methods to separate and analyse potential drug candidates from large compound libraries.
- ☆ Furthermore, methods like liquid-liquid extraction and solid-phase extraction are utilized to isolate and purify drug compounds, facilitating subsequent formulation and preclinical studies.
- ☆ It allows for the separation and identification of active pharmaceutical ingredients, impurities, and degradation products.

4.2. Clinical diagnostics

- ☆ Separation methods also play a critical role in clinical diagnostics, aiding disease detection, monitoring, and treatment.
- ☆ Chromatography methods are employed to separate and analyse biomarkers, hormones, enzymes, and other molecules in patient samples.

4.3. Biotechnology

- ☆ Separation methods are extensively employed in biotechnology and bioengineering for the purification of biomolecules such as proteins, enzymes, and nucleic acids.
- ☆ Methods like affinity chromatography, gel electrophoresis, and ultrafiltration are used to isolate and purify biomolecules for research, diagnostics, and therapeutics.

4.4. Analytical chemistry

- ☆ Separation methods are extensively used in analytical chemistry for sample preparation, identification, and quantification of analytes.
- ☆ Chromatography, electrophoresis, and extraction methods are utilized to separate and analyse complex mixtures in a wide range of industries, including pharmaceuticals, environmental analysis, and forensic sciences.

4.5. Proteomics and genomics

- ☆ Separation methods play a crucial role in proteomics and genomics research, allowing for the analysis and characterization of proteins and genes.

4.6. Environmental analysis

- ☆ Environmental scientists rely on separation methods to analyse pollutants, contaminants, and toxins in air, water, and soil samples.
- ☆ Gas chromatography (GC) and liquid chromatography (LC) are employed to determine the presence and concentration of pollutants, ensuring compliance with environmental regulations.

4.7. Food processing

- ☆ Separation methods are essential in the food industry to ensure food safety, quality, and compliance.
- ☆ Methods like filtration, centrifugation, and chromatography are used to remove impurities, separate components, and detect contaminants such as pesticides, heavy metals, and microbial pathogens.

4.8. Forensic sciences

- ☆ Separation methods are instrumental in forensic analysis, aiding in the identification and analysis of trace evidence.
- ☆ Chromatography methods, such as gas chromatography-mass spectrometry (GC-MS) and high-performance liquid chromatography (HPLC), are used to analyse drugs, toxins, and other compounds found in forensic samples.

4.9. Petrochemical industry

- ☆ Separation methods play a crucial role in the petrochemical industry for refining crude oil and separating hydrocarbons.
- ☆ Distillation is used to separate crude oil into various fractions based on boiling points, while methods like adsorption and membrane filtration are employed for further purification and separation of specific components.

4.10. Water treatment

- ☆ Separation methods are integral to water treatment processes, allowing for the removal of contaminants and impurities.
- ☆ Methods such as sedimentation, filtration, and reverse osmosis are employed to purify water for drinking, industrial use, and wastewater treatment.

Note

- ☆ **Difference between a method and a technique:** In the context of scientific research, the terms "method" and "technique" are often used interchangeably, but they have slightly different meanings:
- ☆ **Technique:** A technique refers to a specific procedure or approach used to perform a particular task or achieve a desired outcome. It typically involves a set of skills, tools, and practices employed in a specific field or discipline. Techniques are often practical and hands-on, focusing on the execution of a process or procedure.

- **Method:** A method, on the other hand, is a broader term that encompasses a systematic and organized approach to solve a problem or answer a research question. It involves a series of steps or procedures designed to achieve a specific objective or obtain reliable results. Methods can incorporate multiple techniques and may include considerations such as sample preparation, data collection, analysis, and interpretation.
- In summary, a technique is a specific procedure or skill used within a method, while a method is a more comprehensive approach that incorporates various techniques and encompasses the overall systematic process followed in a research or experimental study.

02 Chromatography Methods

CHAPTER SUMMARY

- Chromatography methods are powerful analytical techniques used for the separation and identification of compounds in a mixture. They rely on the differential migration of components through a stationary phase.
- The basic principle of chromatography involves the separation of components based on their differential interactions with a stationary phase and a mobile phase, leading to distinct migration rates.
- Chromatographic methods are classified on the basis of stationary and mobile phases or mechanisms of separation or applications.
- Different chromatographic methods offer unique advantages and applications in the separation of biomolecules.
- These methods have applications in diverse fields such as biotechnology, pharmaceuticals, forensics, environmental analysis, and biochemistry, enabling precise separation, identification, and quantification of compounds.

1. Introduction

- The physical and chemical properties can be used to separate the different components of a simple mixture. However, for a complex mixture such as a mixture of proteins or multiple amino acids, more advanced methods are required. Chromatography is an advanced separation and purification method.
- Chromatography is a powerful separation method to separate, identify, and purify the components of a mixture for qualitative and quantitative analysis.
- Chromatography has revolutionized the fields of chemistry, biology, and medicine. It involves separating and analysing complex mixtures of molecules based on their physical and chemical properties, making it an essential tool for research, development, and quality control in various industries.
- In chromatography, the components to be separated are distributed between two phases, one of which is stationary (stationary phase) while the other (the mobile phase) moves in a definite direction.

Note

- **Chroma:** colour; **-graphein:** to write.
- **Chromatography**: An analytical method used to separate a mixture of chemical compounds into their individual components.
- **Chromatogram:** Graph showing the result of separating the components of a mixture by chromatography.

2. Historical Background

- The history of chromatography dates back to the early 1900s when Russian botanist Mikhail Tsvet used a column of calcium carbonate to separate plant pigments based on their colour. He named the method "chromatography" from the Greek words "chroma", meaning colour and "graphein", meaning to write. Although Tsvet's work was largely ignored by the scientific community at the time, his discovery laid the foundation for modern chromatography.
- The development of chromatography accelerated in the 1930s and 1940s with the introduction of paper and thin-layer chromatography. In paper chromatography, the sample is applied to a strip of filter paper and then allowed to migrate up the paper using a solvent. Different components of the sample move at different rates, allowing for their separation and identification. Thin-layer chromatography is a similar method but uses a thin layer of silica gel or alumina on a glass plate as the stationary phase.
- The advent of gas chromatography in the 1950s marked a significant milestone in the history of chromatography. Gas chromatography involves separating volatile compounds based on their partitioning between a stationary phase and a mobile phase. The development of capillary columns and the use of packed columns with temperature programming greatly improved the efficiency and resolution of gas chromatography.
- Liquid chromatography was also introduced in the 1950s and has become one of the most widely used chromatographic methods. It involves separating non-volatile and polar compounds based on their interactions with a stationary phase and a mobile phase. The use of high-performance liquid chromatography (HPLC) with modern column packing materials, detectors, and computerized data analysis has made liquid chromatography a highly versatile and efficient method for separating and quantifying complex mixtures of molecules.
- In the 1970s and 1980s, advances in technology and materials led to the development of new chromatographic methods such as ion chromatography,

size exclusion chromatography, and affinity chromatography. Ion chromatography involves separating ions based on their interactions with a stationary phase that has ion exchange properties. Size exclusion chromatography separates molecules based on their size and shape, while affinity chromatography separates molecules based on their specific interactions with a ligand attached to a stationary phase.

☆ In recent years, chromatography has continued to evolve with the introduction of new column materials, detectors, and data analysis tools. The use of high-resolution mass spectrometry with chromatography has greatly expanded the scope of chromatography, allowing for the identification and quantification of trace amounts of molecules in complex mixtures.

☆ The historical development of chromatography has been marked by a series of breakthroughs and innovations that have transformed the field of analytical chemistry. From the early work of Tsvet to the modern-day applications of high-resolution chromatography-mass spectrometry, chromatography has played a critical role in advancing our understanding of the composition and properties of complex mixtures of molecules.

3. Basic Principles of Chromatographic Methods

☆ The principle of chromatography is based on the fact that when molecules in a mixture are applied onto a surface or into a solid or fluid stationary phase (matrix), they separate from each other while moving with the help of a mobile phase.

☆ In most chromatography methods, phase A is the stationary phase or matrix, and phase B is the mobile phase or buffer.

☆ Chromatographic methods are based on the molecular properties of compounds related to adsorption (liquid-solid), partition (liquid-solid), affinity, or differences among their molecular weights.

☆ Because of differences in molecular characteristics, certain components of the mixture remain in the stationary phase for a longer period of time and travel slowly through the chromatographic system, while others flow rapidly into the mobile phase and exit the system faster.

☆ Thus, all chromatographic methods are based on stationary and mobile phases.

3.1. Stationary phase

☆ The stationary phase is a solid or liquid particle attached to a solid support (glass or a metal surface). In the stationary phase, the components of the mixture to be separated are absorbed selectively.

- The term stationary refers to the fact that this phase remains stationary (fixed) while the other phase moves down the chromatographic system.
- Components of the mixture with similar properties to the stationary phase elute later.
- Most substances used as stationary phases are porous, thereby allowing the attachment of components during chromatography.
- The substance used for the stationary phase of a chromatography method is selected based on the properties of the components to be separated and the type of chromatography.
- Paper, silica, glass, gel beads, gases, or liquids are the most common stationary phase substances.
- Based on the type of stationary phase, the chromatographic methods are divided into column chromatography and planer chromatography.

3.2. Mobile phase

- The mobile phase is either liquid or gas that is passed through a chromatographic system that moves the components of the mixture to be separated at different rates by adsorbing to the stationary phase.
- The mobile phase is the solvent that carries the mixture (containing the components) as it moves down the stationary phase.
- The term mobile refers to the fact that the phase moves down the chromatographic system, whereas the other phase remains stationary.
- Components of the mixture with similar properties to the mobile phase elute earlier.
- The substance used for the mobile phases of a chromatography method is selected based on the properties of the components to be separated and the type of chromatography.
- The most common mobile phase substances are water, alcohol, acetic acid, acetone, or some gases.
- Based on the type of mobile phase, the chromatographic methods are divided into liquid chromatography and gas chromatography.

Note

- **Liquid chromatography:** Chromatography methods that separate compounds in a liquid mobile phase using a solid stationary phase. It is carried out either in a column or a plane. The ion exchange, gel filtration, affinity, hydrophobic, reverse phase, HPLC, paper, and thin-layer chromatography are types of liquid chromatography.
- **Gas chromatography:** Chromatography methods that separate compounds in a gaseous or liquid sample in a gaseous mobile phase (called carrier gas) by passing the gas through a stationary phase. The gas-solid and gas-liquid chromatography are types of gas chromatography.
- The interaction between the stationary phase, mobile phase, and components present in the mixture regulates the separation of compounds from each other.

4. Classification of Chromatographic Methods

- Chromatography methods can be classified based on various criteria, such as the nature of the stationary/mobile phase, mechanism of separation, or intended application.
- The classification of chromatography methods allows scientists to choose the most appropriate method for their specific research needs.
- The major classifications of chromatographic methods are discussed below:

4.1. Based on stationary and mobile phases

4.1.1. Gas chromatography

- Gas chromatography (GC) uses a gas as the mobile phase and a solid or liquid stationary phase.
- The mobile phase is usually an inert gas, and the stationary phase is coated onto a solid support.
- GC is used to separate volatile compounds such as organic compounds, amino acids, and fatty acids.

4.1.2. Liquid chromatography

- Liquid chromatography (LC) uses a liquid mobile phase and a solid/ liquid stationary phase. The stationary phase is usually a solid support, and the mobile phase can be a single solvent or a mixture of solvents.
- The components of the sample mixture are separated based on their relative solubilities in the mobile and stationary phases.

- LC is used to separate a wide range of compounds, including peptides, proteins, nucleic acids, and carbohydrates.

4.2. Based on the mechanism of separation

4.2.1. Adsorption chromatography

- Adsorption chromatography includes chromatography methods that are carried out in a planar or flatbed and the mobile phase moves by capillary action.
- The thin-layer and paper chromatography are types of adsorption chromatography.
- These methods use a solid adsorbent to separate compounds based on their affinity for the adsorbent. The compounds are dissolved in a solvent and passed through the adsorbent. The compounds bind to the adsorbent, and the different compounds have different affinities for the adsorbent. This allows the compounds to be separated.
- The components of the mixture are separated based on their relative affinities to the stationary and mobile phases.
- Adsorption chromatography is used to separate a wide range of compounds, including amino acids, peptides, proteins, and nucleic acids.

4.2.2. Column chromatography

- Column chromatography methods are carried out in a glass or metal column.
- Here, the sample mixture is loaded onto a vertical column containing a stationary phase. A solvent or mixture of solvents, called the mobile phase, is then passed through the column. As the mobile phase moves through the stationary phase, different components of the sample mixture interact differently with the stationary phase and mobile phase and thus elute at different times. The eluted components are collected and analysed further.
- The ion exchange, gel filtration, affinity, hydrophobic, reverse phase, and HPLC are types of column chromatography.

4.3. Based on the application

4.3.1. Analytical chromatography

- Analytical chromatography focuses on identifying and quantifying the components of a sample.
- It is a powerful analytical tool used in various fields such as pharmaceuticals, food science, environmental analysis, and forensics.

- ☆ The primary objective of analytical chromatography is to separate the components of a mixture and to detect and quantify them.
- ☆ Analytical chromatography typically involves the use of high-performance liquid chromatography (HPLC) or gas chromatography (GC), and the separated components are analysed using a detector.
- ☆ The information obtained from analytical chromatography is used to determine the composition of the sample, evaluate the purity of the compounds, and identify unknown components.

4.3.2. Preparative chromatography

- ☆ Preparative chromatography is used to isolate and purify specific compounds from a mixture.
- ☆ It involves the use of a stationary phase and a mobile phase to separate the desired compound from other components in the mixture.
- ☆ It is widely used in the pharmaceutical industry to produce pure drugs and in the production of natural products such as plant extracts. The method is also used in research laboratories to isolate specific compounds for further study.
- ☆ The primary objective of preparative chromatography is to isolate and purify specific compounds from a mixture for further analysis or use.
- ☆ Preparative chromatography typically uses larger quantities of sample than analytical chromatography and can be performed using various types of chromatography such as column chromatography, HPLC, or flash chromatography.

5. Types of Chromatographic Methods

- ☆ In chromatography, compounds are separated based on their properties, such as size, shape, charge, hydrophobic groups present on the surface, and binding capacity with the stationary phase. Hence, there are various chromatography methods, each with its own instrumentation and working principle.
- ☆ Based on the classification of stationary and mobile phases, the common chromatography methods are listed below **(Table 1).**

General classification	Specific chromatography method	Stationary phase	Mobile phase	Type of equilibrium
Liquid chromatography	Ion-exchange	Ion-exchange resin with ionized groups linked to a solid surface	Liquid	Ion-exchange
	Gel filtration	Liquid in the interstices of a polymeric solid matrix	Liquid	Partition/ Sieving
	Affinity	Group-specific liquid linked to a polymeric solid surface	Liquid	Partition
	Hydrophobic interaction	Inert matrix of spherical particles coated with ligands containing alkyl or aryl groups	Liquid containing high salts	Adsorption
	Reverse phase	Silica beads covered with carbon chains or naked hydrophobic polymers	Very polar liquid containing water-miscible solvents	Partition
	HPLC	Silica-based or polymer-based solid matrix	Liquid	Based on chromatography method
	Paper	Solid	Liquid	Adsorption
	Thin-layer	Solid	Liquid	Adsorption
Gas chromatography	Gas-solid	Solid	Gas	Adsorption
	Gas-liquid	Liquid adsorbed or bonded to a solid surface	Gas	Partition between gas and liquid

6. Applications of Chromatography

- ☆ Chromatography is a versatile analytical method that has a wide range of applications in various industries, including pharmaceuticals, biotechnology, food and beverage, environmental analysis, and forensic science.
- ☆ Its ability to separate and identify complex mixtures of molecules makes it an essential tool for research, development, and quality control in various industries. With continued advances in technology and materials, chromatography is likely to play an increasingly important role in the future of analytical chemistry.

☆ Some of the key applications of chromatography are discussed below.

6.1. Pharmaceutical and chemical industry

☆ Chromatography is a critical tool in the development and quality control of pharmaceuticals. It is used to separate and identify active ingredients, impurities, and degradation products in drug formulations.

☆ High-performance liquid chromatography (HPLC) and gas chromatography (GC) are commonly used in pharmaceutical analysis.

☆ Chromatographic methods are used in various stages of drug development, including the preparation and purification of drugs.

6.2. Biotechnology

☆ Chromatography plays a key role in the purification and analysis of biologically derived products such as proteins, antibodies, and nucleic acids.

☆ Affinity chromatography, ion exchange chromatography, and size exclusion chromatography are commonly used in biotechnology applications.

☆ In life sciences, it is used for the purification of proteins, enzymes, nucleic acids, and other biomacromolecules.

☆ In chemistry, it is used to separate mixtures of small molecules and to purify compounds after synthesis.

6.3. Food industry

☆ Chromatography is used in the analysis of food and beverage products to ensure their quality and safety. It is used to detect and quantify contaminants, additives, and flavour compounds.

☆ HPLC and GC are commonly used in food and beverage analysis.

☆ They are also used for the removal of contaminants from water.

6.4. Environmental analysis

☆ Chromatography is used to analyse environmental samples such as soil, water, and air to detect and quantify pollutants and other contaminants.

☆ Gas chromatography-mass spectrometry (GC-MS) is commonly used in environmental analysis.

6.5. Forensic science

☆ Chromatography is used in forensic analysis to detect and quantify drugs, explosives, and other substances in crime scene samples. GC-MS and HPLC are commonly used in forensic analysis.

6.6. Other applications

- ☆ In addition to these applications, chromatography is also used to analyse complex mixtures such as petroleum products, polymers, and natural products. The versatility and sensitivity of chromatography make it an essential tool in modern analytical chemistry.

03 Paper Chromatography

CHAPTER SUMMARY

- Paper chromatography is a simple yet powerful method that is widely used for the separation and analysis of compounds.
- It operates on the principle of differential migration of components in a mixture through a stationary phase (paper) due to varying affinities with the mobile phase (solvent).
- There are several types of paper chromatography methods and each method offers unique advantages and is suitable for different applications in separation and analysis.
- The methodology involves the separation of components in a sample based on their differential migration on a paper strip or sheet using a mobile phase. The method relies on the principles of adsorption, capillary action, and solute partitioning for effective separation and analysis.
- Paper chromatography enables the separation and identification of components in complex mixtures, making it a valuable tool in research and quality control.

1. Introduction

- ☆ Paper chromatography is a type of chromatography that uses paper as a stationary phase and a liquid or gas as a mobile phase to separate mixtures.
- ☆ It is a simple and inexpensive method widely used to separate complex mixtures into their individual components.
- ☆ It has many advantages over other separation methods, such as its simplicity, versatility, and low cost.
- ☆ Paper chromatography was first introduced by Martin and Synge in 1941 and has since become an important separation tool.
- ☆ The basic principle of paper chromatography is the differential migration of components in a mixture due to differences in their solubility and adsorption properties.

☆ Paper chromatography has a wide range of applications in various fields. It is widely used in the chemical industry to separate and analyse organic compounds, including amino acids, peptides, and carbohydrates. It is extensively used in the pharmaceutical industry to analyse drugs and their metabolites. It is also used in food analysis to detect additives and contaminants. In forensic science, it is used to analyse drugs, poisons, and explosives.

☆ The methodology of paper chromatography is simple and requires only a few materials and equipment. The materials required include filter paper, a mobile phase, and a sample to be analysed. The mobile phase can be either a liquid or a gas, depending on the type of chromatography being used. The equipment required includes a glass container or chamber, a spotting device, and a developing tank.

2. Principles

2.1. Basic principles

☆ The basic principle of paper chromatography is the differential migration of components in a mixture due to differences in their solubility and adsorption properties.

☆ The mixture to be separated is applied as a spot or a line on the paper, and the paper is then placed in a container containing a mobile phase (Figure 1). The mobile phase moves up the paper by capillary action, carrying the mixture along with it. The mobile phase can be a liquid or a gas, depending on the type of chromatography being used. In the case of liquid chromatography, the mobile phase is a liquid solvent that moves up the paper by capillary action. In the case of gas chromatography, the mobile phase is a carrier gas that carries the mixture along with it.

☆ As the mobile phase moves up the paper, it takes the individual components of the mixture along with it. The more soluble and less adsorptive components will travel further up the paper, while those that are less soluble and more adsorptive will remain closer to the point of application, thereby separating components (Figure 1). This separation occurs due to the differential partitioning of the components between the stationary phase (the paper) and the mobile phase (the solvent or gas).

☆ The separation of the components can be visualised by the appearance of spots or bands on the paper. Each spot or band corresponds to a different component in the mixture. The spots can be visualised using a UV lamp, a staining agent, or a ninhydrin spray.

☆ The separation of the components is based on various factors, such as the chemical nature of the mixture, the properties of the stationary and mobile phases, and the choice of solvent or gas. The efficiency of the separation can be enhanced by optimising these factors.

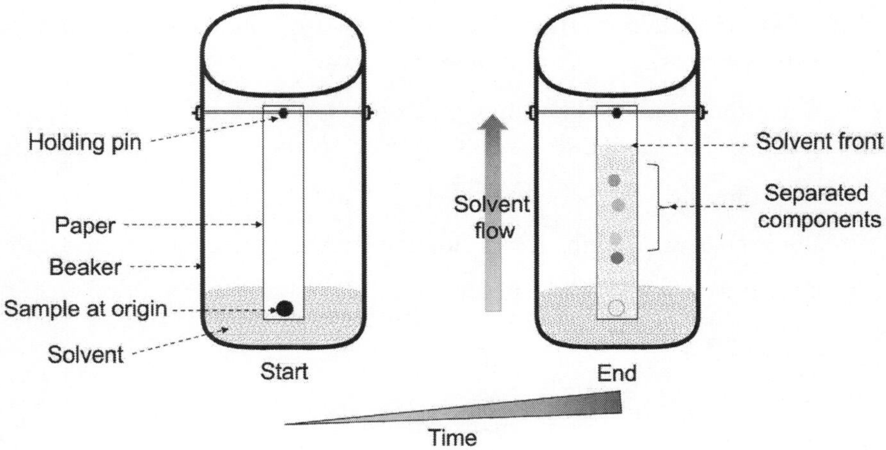

Figure 1: Separation of sample components using paper chromatography. A paper strip is placed in a solvent, and the sample is spotted. The sample ascends and separates the mixture into its components. The stationary phase (paper) and the mobile phase (solvent) allow for the distinct migration of each component.

2.2. Interactions used in paper chromatography

☆ Paper Chromatography relies on various interactions between the components of the mixture and the stationary and mobile phases to achieve separation.

☆ The choice of interaction used in paper chromatography depends on the nature of the mixture being analysed and the specific requirements of the separation. Each type of interaction has its advantages and disadvantages, and the optimal choice depends on the specific needs of the experiment.

☆ The interactions used in paper chromatography include adsorption, partition, and ion exchange.

2.2.1. Adsorption

☆ Adsorption is the most commonly used interaction in paper chromatography.

☆ In this type of interaction, the components of the mixture are separated based on their differential affinity for the stationary phase (paper) and the mobile phase (solvent or gas).

- ☆ The components that are more strongly adsorbed to the stationary phase will move more slowly than those that are less strongly adsorbed.

2.2.2. Partition

- ☆ Partition is another type of interaction used in paper chromatography.
- ☆ This type of interaction involves the differential solubility of the components between the stationary and mobile phases.
- ☆ The components that are more soluble in the mobile phase will move more quickly than those that are less soluble.

2.2.3. Ion-exchange

- ☆ Ion-exchange is a third type of interaction used in paper chromatography.
- ☆ This type of interaction involves the differential attraction of charged components to the stationary phase.
- ☆ The components that are more strongly attracted to the stationary phase will move more slowly than those that are less strongly attracted.

2.3. Factors affecting paper chromatography

- ☆ The performance of paper chromatography can be influenced by a range of factors, including the solvents used, the choice of stationary and mobile phases, the properties of the sample being analysed, and the experimental conditions used.
- ☆ Understanding and optimising these factors is crucial for achieving reliable and accurate results using paper chromatography.

2.3.1. Stationary phase

- ☆ One of the most important factors affecting paper chromatography performance is the choice of stationary phase.
- ☆ The stationary phase is the surface on which the sample components are separated.
- ☆ In paper chromatography, the stationary phase is typically a piece of paper or cellulose.
- ☆ The properties of the stationary phase, such as pore size and surface area, can significantly affect the separation efficiency of the method.

2.3.2. Solvents

- ☆ Solvents are used to dissolve the sample being analysed and to create the mobile phase.
- ☆ The choice of solvent is critical, as it determines the polarity and chemical properties of the mobile phase.

- ☆ The solvent must be miscible with the sample being analysed and have a suitable boiling point to prevent evaporation during the separation process.
- ☆ In addition, the solvent should be inert and not react with the sample or the stationary phase.

2.3.3. Mobile phase

- ☆ The mobile phase moves through the stationary phase and takes the sample components with it.
- ☆ The mobile phase must be able to dissolve the sample components and must be selected based on the properties of the sample being analysed.
- ☆ The choice of the mobile phase is critical as it determines the interactions between the sample components and the stationary phase, which ultimately affects the separation efficiency.
- ☆ Factors such as polarity, viscosity, and pH can all affect the separation efficiency of the method.

2.3.4. Sample

- ☆ The properties of the sample being analysed can also influence the performance of paper chromatography.
- ☆ Factors such as the size, shape, and chemical nature of the sample components can all affect their retention on the stationary phase and hence the separation efficiency of the method.

2.3.5. Experimental conditions

- ☆ The experimental conditions used can also significantly impact the performance of paper chromatography.
- ☆ Factors such as temperature, pressure, and the flow rate of the mobile phase can all affect the separation efficiency of the method.

2.3.6. Interactions between the components of the mixture and the stationary and mobile phases

- ☆ The interactions between the components of the mixture and the stationary and mobile phases are essential for the success of paper chromatography.
- ☆ Depending on their polarity and chemical properties, the sample components interact differently with the stationary and mobile phases.
- ☆ More polar components interact more strongly with the stationary phase and move more slowly through the mobile phase, while less polar components move more quickly.

2.4. Retention factor (Rf value)

- The retention factor (Rf value) is a crucial parameter in TLC used to identify and quantify components in a mixture.
- By comparing the Rf values of unknown compounds with those of known standards, the identity of the compounds in a mixture can be determined.
- The Rf value is a measure of the distance travelled by a component to the distance travelled by the solvent front.
- Rf values are useful in the interpretation of TLC results and can be used to compare the behaviour of different compounds in the same system.
- Rf values can also be used to identify unknown compounds by comparing their Rf values to those of known compounds.

3. Types of Paper Chromatography

- Paper chromatography can be divided into several types based on the differences in the stationary phase, mobile phase, and mode of separation. The most commonly used types of paper chromatography include:

3.1. Ascending paper chromatography

- In this type of chromatography, the stationary phase is a piece of paper, and the mobile phase moves up the paper through capillary action.
- This method is useful for separating small molecules, such as amino acids and nucleotides.
- In this method, a small spot of the sample is placed on a strip of filter paper at the bottom, known as the origin, and the paper is placed in a solvent. The solvent wicks up the paper, and as it moves, it takes the different components of the sample with it. The components of the sample separate out based on their affinity for the stationary phase and the mobile phase and their physical and chemical properties.
- The separated components can be visualised using staining, fluorescence, and UV absorption.

3.2. Descending paper chromatography

- This type of chromatography is similar to ascending paper chromatography, except that the mobile phase moves down the paper due to gravity.
- This method is useful for separating large molecules, such as proteins and carbohydrates.
- Here, the paper is hung vertically, and the solvent is added to the top. The sample is placed at the top of the paper, and the solvent moves downwards, carrying the components of the sample with it.

- ☆ Descending paper chromatography is often used when the components of the sample are more soluble in the solvent than in water, as the solvent moves down the paper more quickly than the water.

3.3. Two-dimensional paper chromatography

- ☆ In this method, the paper is first run in one direction, then rotated 90 degrees, and run in a second direction.
- ☆ Two-dimensional paper chromatography allows for better separation of components in a sample, as the sample is exposed to different solvent systems in each direction.

3.4. Circular paper chromatography

- ☆ In this method, a circular piece of paper is used as the stationary phase, and the mobile phase is applied to the centre of the circle.
- ☆ As the mobile phase moves outward, it takes the components of the sample with it.
- ☆ Circular paper chromatography is useful for separating components of a sample that have similar physical and chemical properties.

3.5. Preparative paper chromatography

- ☆ This method is used for large-scale separation and purification of samples.
- ☆ It involves using larger sheets of paper and larger amounts of mobile and stationary phases.

3.6. Reversed-phase paper chromatography

- ☆ In this type of chromatography, a hydrophobic stationary phase is used, and the mobile phase is polar.
- ☆ This method is useful for separating non-polar molecules, such as lipids and steroids.

4. Methodology

- ☆ The procedure of paper chromatography involves several steps, including the preparation of the sample, preparation of the paper, selection of the solvent system, and running of the sample.

1. **Preparation of the sample:** The first step in paper chromatography is the preparation of the sample. The sample must be dissolved in a suitable solvent to create a homogeneous solution. The sample should be placed on the paper in a small spot, usually near the bottom of the paper. The amount of sample used will depend on the sensitivity of the method and the quantity of the sample available.

2. **Preparation of the paper:** The second step in paper chromatography is the preparation of the paper. The paper should be of high quality and have a uniform thickness. The paper should be cut to the appropriate size and shape and marked with a pencil to indicate the origin and position of the sample. The paper should be placed in a suitable chamber, such as a glass jar, and allowed to equilibrate with the solvent system.
3. **Selection of the solvent system:** The selection of the solvent system is a critical step in paper chromatography. The solvent system should be selected based on the properties of the sample and the desired separation. The solvent should be volatile and not react with the sample or the paper. The solvent should have a suitable polarity and viscosity to ensure adequate separation of the components of the sample.
4. **Running of the sample:** Once the paper has been prepared and the solvent system selected, the paper is placed in the chamber, and the solvent is added to the bottom of the chamber. The solvent moves up the paper, carrying the components of the sample with it. The separation of the components is based on their affinity for the stationary phase (the paper) and the mobile phase (the solvent). The different components of the sample will travel different distances up the paper, depending on their physical and chemical properties.

Note

- ☆ The separated components can be visualised using various methods, including staining, fluorescence, or UV absorption. The components can also be eluted from the paper and further analysed using other analytical methods, such as spectroscopy or mass spectrometry.

5. Applications

- ☆ Paper chromatography is a widely used analytical method with numerous applications in various fields, including chemistry, biology, pharmacology, and environmental science.
- ☆ The method is particularly useful for separating and identifying complex mixtures of compounds and has the advantage of being simple, inexpensive, and easy to perform.
- ☆ It can be used to separate and identify complex mixtures of compounds, purify proteins, detect environmental pollutants, and monitor the quality of food and beverages. As a result of its simplicity, low cost, and wide applicability, paper chromatography is likely to remain an important tool for researchers and analysts in a variety of fields.

☆ Below are some of the most common applications of paper chromatography:

5.1. Qualitative analysis of organic compounds

☆ Paper chromatography is frequently used to identify and characterise organic compounds such as amino acids, sugars, and vitamins.

☆ By comparing the Rf values of unknown compounds with those of known standards, the identity of the compounds in a mixture can be determined.

5.2. Quantitative analysis of drugs and metabolites

☆ Paper chromatography is an important tool in pharmacology and toxicology, where it is used to measure the concentrations of drugs and their metabolites in biological samples such as blood, urine, and tissue.

☆ The method can also be used to detect and quantify environmental pollutants.

5.3. Separation of plant pigments

☆ Paper chromatography is widely used in plant biology to separate and identify the pigments present in leaves, flowers, and fruits. By using different solvent systems and chromatography papers, researchers can obtain highly selective and reproducible separations.

5.4. Protein purification

☆ Paper chromatography is used to purify and fractionate proteins based on their physicochemical properties, such as charge, size, and hydrophobicity.

☆ By combining different chromatography steps, researchers can obtain highly purified proteins for further analysis.

5.5. Forensic analysis

☆ Paper chromatography is frequently used in forensic science to analyse evidence such as ink, dyes, and fibres.

☆ By comparing the chromatographic profiles of samples from different sources, investigators can determine whether the samples came from the same source.

5.6. Quality control of food and beverages

☆ Paper chromatography is used in the food and beverage industry to monitor the quality and purity of products such as wine, beer, and fruit juices.

☆ By analysing the chromatographic profiles of different batches of products, manufacturers can ensure consistency and identify any contaminants or adulterants.

04 Thin-Layer Chromatography

CHAPTER SUMMARY

- Thin-layer chromatography (TLC) is a versatile separation method that uses a thin layer of stationary phase to separate and analyse compounds based on their different affinities and interactions.
- The principle is based on the differential migration of components in a sample on a thin layer of stationary phase (such as silica gel or cellulose), driven by capillary action and their varying affinities to the stationary phase.
- There are several types of TLC methods, which differ in the polarity of the stationary phase and mobile phase interactions.
- The methodology of TLC involves the application of the sample onto a thin adsorbent layer and subsequent separation using a mobile phase.
- TLC is widely used in the analysis of pharmaceuticals, pesticides, natural products, dyes, and pigments, as well as for the identification of unknown substances and separation of complex mixtures.

1. Introduction

- Thin-layer chromatography (TLC) is a widely used separation method that is based on the principles of adsorption chromatography.
- It allows for the separation, identification and quantification of various compounds in a sample.
- TLC is a rapid, cost-effective, and simple method that requires minimal sample preparation and equipment.
- The stationary phase is usually a thin layer of an adsorbent material coated onto a flat support, while the mobile phase is a solvent. The sample mixture is applied to the surface of the stationary phase and is allowed to dry. The plate is then placed in a developing chamber, and the solvent migrates up the plate by capillary action.

- ☆ The separation of the components is based on their affinity for the stationary phase and the mobile phase. Components with a higher affinity for the stationary phase will move more slowly up the plate, while those with a higher affinity for the mobile phase will move faster. This results in the formation of distinct bands or spots on the plate, each corresponding to a different component of the sample mixture.
- ☆ TLC has several advantages over other separation methods, such as column chromatography, gas chromatography, or high-performance liquid chromatography. It requires minimal sample preparation, can be performed quickly, and does not require expensive equipment. Additionally, it is useful for the separation and identification of non-volatile compounds that cannot be analysed by gas chromatography.
- ☆ TLC is widely used in the chemical, biological, and pharmaceutical sciences to analyse complex mixtures, identify unknown compounds, and purify compounds. In the biological sciences, it is used to separate proteins, nucleic acids, and lipids. It is also used in forensic science, food analysis, and environmental monitoring. For example, TLC is used to analyse drugs, vitamins, fatty acids, and amino acids.

2. Principles

2.1. Basic principles

- ☆ TLC is based on the principles of adsorption chromatography, which involve the separation of the components of a sample mixture based on their interactions with a stationary phase and a mobile phase.
- ☆ In TLC, the stationary phase is a thin layer of an adsorbent material, such as silica gel or alumina, that is coated onto a flat support, while the mobile phase is a solvent or mixture of solvents.
- ☆ The principle of TLC is based on the fact that different compounds have different affinities for the stationary and mobile phases, and this results in their separation into individual bands or spots on the plate.
- ☆ When a sample mixture is applied to the surface of the stationary phase, the components of the mixture interact with the stationary phase. The components with a higher affinity for the stationary phase move slowly up the plate, while those with a higher affinity for the mobile phase move faster. As the mobile phase migrates up the plate by capillary action, the different components of the mixture separate into distinct bands or spots, each corresponding to a different component of the sample mixture (Figure 1). The separation is based on the relative polarity, size, and shape of the molecules in the mixture, as well as the type of stationary and mobile phases used.

☆ By comparing the Rf values of unknown compounds with those of known standards, the identity of the compounds in a mixture can be determined (*for details on Rf values, see the chapter on paper chromatography*).

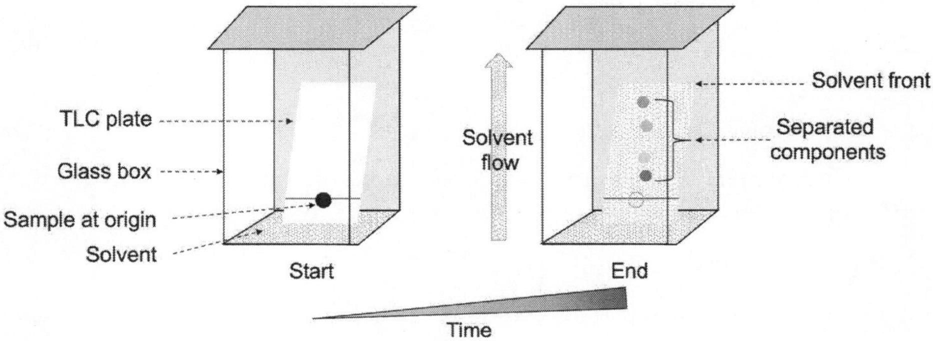

Figure 1: Thin-layer chromatography of a sample mixture on a silica gel plate. The plate is first spotted with the sample and then developed in a solvent using capillary action. After the solvent front reaches the top of the plate, it is removed from the chamber and visualized. The different pigments are separated based on their polarity and Rf values.

2.2. Interactions used in TLC

☆ In TLC, the components are separated based on their interactions with a stationary phase and a mobile phase. The interactions between the components of the mixture, the stationary phase, and the mobile phase are dependent on the physical and chemical properties of the molecules involved.

☆ The most common types of interactions used in TLC include:

2.2.1. Van der Waals interactions

☆ These are weak, non-polar interactions that occur between non-polar molecules and non-polar regions of polar molecules.

☆ Van der Waals interactions are important in the separation of non-polar compounds in TLC.

2.2.2. Dipole-dipole interactions

☆ These are stronger than Van der Waals interactions and occur between polar molecules or polar regions of molecules.

☆ Dipole-dipole interactions are important in the separation of polar compounds in TLC.

2.2.3. Hydrogen bonding

- This is a strong intermolecular force that occurs when a hydrogen atom is covalently bonded to a highly electronegative atom, such as oxygen or nitrogen.
- Hydrogen bonding is important in the separation of compounds containing functional groups, such as alcohols and carboxylic acids.

2.2.4. Ion-dipole interactions

- These occur between ions and polar molecules and are important in the separation of charged compounds in TLC.

2.3. Factors affecting TLC

- The performance of TLC can be influenced by several factors that affect the efficiency, selectivity, and accuracy of the separation.
- The factors include the choice of stationary phase and mobile phase, sample preparation, and the experimental conditions during the separation. Some of the factors that can affect TLC performance are:

2.3.1. Choice of stationary phase

- One of the most crucial factors in TLC is the choice of stationary phase.
- The choice of stationary phase is based on the polarity and chemical properties of the analytes to be separated and can affect the separation efficiency and selectivity of TLC.
- Different types of adsorbents have different properties, such as particle size, surface area, and pore size, which can affect the interaction between the stationary phase and the sample components. For example, small particle size and high surface area will allow for more efficient separation.
- Several types of stationary phases are used in TLC, including silica gel, alumina, cellulose, and polyamide.
  - **Silica gel** is the most common stationary phase used in TLC and is suitable for separating a wide range of compounds.
  - **Alumina** is typically used for compounds that are difficult to separate on silica gel, while cellulose is used to separate carbohydrates and other polar compounds.
  - **Polyamide** is used to separate polar compounds that are not well separated on silica gel or cellulose.

2.3.2. Choice of mobile phase

- The choice of the mobile phase can also affect the separation efficiency and selectivity of TLC.

- ☆ The mobile phase should have sufficient solvent strength to dissolve the sample components and move them along the stationary phase. It should also have the appropriate polarity to separate the sample components effectively.
- ☆ The polarity, viscosity, and volatility of the mobile phase can all affect the migration and separation of the sample components.
- ☆ Several types of solvent systems are used in TLC, including non-polar solvents, moderately polar solvents, and polar solvents.
  - ☆ **Non-polar solvents** such as hexane and heptane are commonly used for separating non-polar compounds.
  - ☆ **Moderately polar solvents** such as ethyl acetate and chloroform are used for separating moderately polar compounds.
  - ☆ **Polar solvents** such as methanol and water are used for separating polar compounds.
- ☆ In addition to these solvent types, there are also mixed solvent systems that combine solvents from different categories to achieve a desired separation. For example, a mixture of ethyl acetate and hexane can be used to separate compounds with intermediate polarity.

2.3.3. Sample preparation

- ☆ Proper sample preparation is crucial for achieving good separation results in TLC.
- ☆ Factors such as the amount of sample applied, the solvent used for dissolving the sample, and the purity of the sample can all affect the separation results.
- ☆ A high concentration of the sample can lead to overloading, resulting in poor separation. Impurities in the sample can also interfere with the separation by adsorbing onto the stationary phase, leading to peak tailing or broadening.

2.3.4. Experimental conditions

- ☆ Factors such as temperature, humidity, and the duration of the separation can also affect the performance of TLC. These factors can affect the mobility of the sample components, leading to changes in the separation efficiency and selectivity.

3. Types of TLC

- ☆ There are several types of TLC methods, each with its own unique properties and applications.

- ☆ The choice of the TLC method depends on the polarity, size, and chemical properties of the compounds to be separated.
- ☆ The type of stationary phase and mobile phase used in TLC are critical for achieving good separation and accurate identification of compounds.

3.1. Normal phase TLC

- ☆ Here, the stationary phase is polar, such as silica gel or cellulose, and the mobile phase is non-polar, such as hexane or chloroform.
- ☆ This type of TLC is suitable for separating non-polar to moderately polar compounds.
- ☆ Normal phase TLC is commonly used to separate and analyse lipids, steroids, and alkaloids.

3.2. Reversed-phase TLC

- ☆ Here, the stationary phase is non-polar, such as C18-bonded silica, and the mobile phase is polar, such as methanol or water.
- ☆ This type of TLC is suitable for separating polar to moderately polar compounds.
- ☆ Reversed-phase TLC is commonly used to separate and analyse peptides, nucleic acids, and other polar compounds.

3.3. Chiral TLC

- ☆ Here, the stationary phase contains a chiral compound, like cyclodextrin.
- ☆ This type of TLC is suitable for separating enantiomers of chiral compounds.
- ☆ Chiral TLC is commonly used to separate and analyse pharmaceuticals, natural products, and other chiral compounds.

3.4. Ion-exchange TLC

- ☆ Here, the stationary phase is an ion-exchange resin, and the mobile phase contains ions of opposite charge.
- ☆ This type of TLC is suitable for separating charged compounds such as amino acids or peptides.
- ☆ Ion-exchange TLC is commonly used in the separation and analysis of biomolecules.

3.5. Size exclusion TLC

- ☆ Here, the stationary phase is a porous polymer or gel, and the separation is based on the size of the compounds.
- ☆ This type of TLC is suitable for separating polymers or biomolecules.

- ☆ Size exclusion TLC is commonly used to separate and analyse proteins, nucleic acids, and carbohydrates.

4. Methodology

- ☆ The procedure of TLC is relatively simple, and the method is widely accessible to researchers and scientists in various fields.
- ☆ The methodology of TLC involves several steps, including sample preparation, plate preparation, chromatography, visualization, and identification.

1. **Sample preparation:** The sample must be dissolved in a solvent that is compatible with the TLC system. The sample must also be filtered to remove any particulate matter that may interfere with the separation.
2. **Plate preparation:** The TLC plate is the stationary phase and is typically made of silica gel or alumina. The plate is coated with the stationary phase and allowed to dry. The plate is then activated by heating in an oven or a fume hood.
3. **Chromatography run:** The prepared plate is placed in a developing chamber with a mobile phase that is compatible with the stationary phase. The mobile phase moves up the plate, carrying the sample with it. The separation occurs as the different compounds in the sample have varying affinities to the stationary phase and mobile phase.
4. **Visualization**: Once the separation is complete, the plate is removed from the developing chamber and allowed to dry. The separated compounds are then visualized using a variety of techniques, such as UV light, iodine vapour, or chemical sprays.
5. **Identification:** Once the compounds are visualized, their Rf values can be calculated. These values can be compared to known standards or literature values to identify the compounds in the sample.

Note:

Comparison between paper and thin-layer chromatography:

- ☆ The main difference between paper chromatography and TLC is the nature of the stationary phase. In paper chromatography, the stationary phase is a piece of filter paper, while in TLC, the stationary phase is a thin layer of silica gel or alumina coated on a glass or plastic plate.
- ☆ Another difference is the level of resolution that can be achieved. TLC is generally more precise and can separate components with higher resolution than paper chromatography.

- ☆ Additionally, TLC is a faster method than paper chromatography and requires less solvent, making it more efficient and cost-effective.

5. Applications

- ☆ TLC is a versatile and valuable analytical method that has a wide range of applications in various fields.
- ☆ TLC has several advantages over other methods, such as high sensitivity, low cost, and rapid analysis. It is also a relatively simple method that can be performed in a laboratory or in the field.
- ☆ However, there are also some limitations to the method. For example, TLC may not be suitable for the analysis of complex mixtures, and it may not be able to separate compounds that have similar physical and chemical properties.
- ☆ With its many advantages and applications, TLC has emerged as an important tool for researchers, scientists, and analysts in a variety of fields.

5.1. Pharmaceuticals

- ☆ One of the most significant applications of TLC is in the pharmaceutical industry.
- ☆ It is commonly used for the analysis of drugs and their impurities.
- ☆ It is also used to monitor the stability of drugs and to assess the quality of raw materials used in drug manufacturing.
- ☆ TLC can rapidly and reliably detect a range of impurities in drugs, including related substances, degradation products, and other impurities that can affect drug quality and safety.

5.2. Research and academia

- ☆ TLC is a rapid, inexpensive, and versatile method that can provide valuable information in a variety of research settings.
- ☆ TLC is widely used for the separation, identification, and quantification of organic and inorganic compounds.
- ☆ In life sciences research, TLC is used to analyse biomolecules such as amino acids, lipids, and carbohydrates.
- ☆ In chemistry, TLC is commonly used to analyse chemical reactions, monitor the purity of a compound, and identify unknown compounds.

5.3. Food analysis

- ☆ TLC is used to detect food additives, such as colourants, preservatives, and sweeteners.

- ☆ It is also used to identify contaminants, such as pesticides and mycotoxins.
- ☆ TLC can detect a wide range of food contaminants and is often used as a screening tool to identify samples that require further analysis.

5.4. Environmental monitoring

- ☆ TLC is used in environmental monitoring to analyse environmental samples, such as soil, water, and air, for pollutants and contaminants.
- ☆ It is also used to monitor the degradation of pollutants over time and to assess the effectiveness of remediation methods.

5.5. Forensics

- ☆ TLC is used in forensic analysis to identify drugs of abuse, explosives, and other illicit substances.
- ☆ It is also used to analyse trace evidence, such as fibres and paint chips, in criminal investigations.

5.6. Natural products

- ☆ TLC is used to analyse the composition of natural products, such as essential oils, plant extracts, and herbal medicines.
- ☆ It is often used to identify the active ingredients in natural products, as well as to detect any impurities or contaminants.

5.7. Other applications

- ☆ TLC has many other applications, including the detection of toxic substances in industrial processes and the analysis of complex mixtures.
- ☆ TLC is often used as a preliminary screening tool to identify samples that require further analysis. It is also used to assess the purity of a sample or to identify the presence of a specific compound.

05 Ion-Exchange Chromatography

CHAPTER SUMMARY

- Ion-exchange chromatography (IEC) is a powerful separation method used to separate and purify charged molecules based on their interactions with charged stationary phases.
- IEC relies on the selective binding of ions to charged functional groups on the stationary phase, allowing for separation based on their charge and affinity.
- Types of IEC include cation-exchange chromatography, which separates cations based on their charge, and anion-exchange chromatography, which separates anions based on their charge.
- In IEC, the sample is loaded onto a column packed with ion-exchange resin. The target ions interact with the resin, while others pass through. Elution is achieved by changing the buffer conditions.
- The performance of IEC is influenced by factors such as pH, ionic strength, temperature, choice of stationary and mobile phases, and sample composition.
- IEC finds applications in various fields including pharmaceutical analysis, protein purification, water treatment, environmental analysis, and separation of charged biomolecules in biological research.

1. Introduction

- Ion-exchange chromatography (IEC) is a valuable separation technique that operates on reversible electrostatic interactions between charged analyte molecules and a solid stationary phase.
- It is used to separate and purify charged molecules based on their affinity for ion-exchange resins (solid stationary phase).
- It relies on the reversible exchange of ions between the sample solution and an ion-exchange resin. The resin contains functional groups that carry either positive or negative charges, which interact with the oppositely charged analyte molecules in the sample solution. This electrostatic interaction allows for the selective retention and separation of different analytes.

- ☆ It operates on the principle of electrostatic interactions between the charged analyte molecules and the oppositely charged functional groups on the resin. The sample is loaded onto a column packed with the ion-exchange resin, and a mobile phase (buffer solution) is passed through the column. The analyte ions bind to the resin through ion-exchange interactions, while unbound species pass through the column more quickly. By adjusting the composition and pH of the mobile phase, analytes can be selectively eluted from the resin, leading to their separation.
- ☆ IEC finds extensive applications in various fields, including biotechnology, pharmaceuticals, environmental analysis, and biochemical research. It is commonly used for protein purification, allowing for the separation of target proteins from complex mixtures. This technique is essential for the production of biopharmaceuticals, monoclonal antibodies, and enzymes.

2. Principles of IEC

2.1. Ion-exchange interactions

- ☆ IEC is based on the principle of reversible ion-exchange interactions between charged analyte molecules and a solid stationary phase.
- ☆ The ion-exchange process occurs due to electrostatic forces between the charged functional groups on the stationary phase and the oppositely charged analytes in the sample solution.
- ☆ The interactions can be of two types: cation exchange and anion exchange.
- ☆ In cation exchange, positively charged analytes interact with negatively charged functional groups on the stationary phase, while in anion exchange, negatively charged analytes interact with positively charged functional groups.

2.2. Stationary phase

- ☆ The stationary phase in IEC is typically a solid resin containing functional groups that carry either positive or negative charges. These functional groups are responsible for the ion-exchange interactions with the analyte ions.
- ☆ The choice of the stationary phase depends on the nature of the analyte and the desired separation. Commonly used stationary phases include sulfonic acid (SO^{3-}) groups for cation exchange and quaternary ammonium ($N^+(CH_3)_3$) groups for anion exchange.

2.3. Mobile phase

- ☆ The mobile phase in IEC is a liquid solution, often called the eluent or buffer solution. It flows through the column, carrying the sample and facilitating the separation of analytes.

☆ The composition and pH of the mobile phase play a crucial role in controlling the ion-exchange interactions. By adjusting these parameters, analytes can be selectively retained or eluted from the stationary phase.

2.4. Ion-exchange capacity

☆ The ion exchange capacity of a resin refers to the number of exchangeable ions it can accommodate per unit weight or volume.

☆ It is a measure of the resin's ability to retain and exchange ions during chromatographic separation.

☆ The ion exchange capacity is determined by the type and concentration of functional groups present on the resin. Resins with higher ion exchange capacities can retain a larger number of ions and provide better separation efficiency.

2.5. Types of ion-exchange resins

☆ There are two main types of ion-exchange resins used in chromatography: cation-exchange resins and anion-exchange resins.

☆ Both cation-exchange and anion-exchange resins are available in various forms, such as beads, powders, or monolithic structures, to accommodate different chromatographic setups and requirements.

2.5.1. Cation-exchange resins

☆ Cation-exchange resins (cation exchangers) have negatively charged functional groups that selectively retain positively charged analytes (Table 1 and Figure 1).

☆ These resins are typically composed of polystyrene divinylbenzene (PS-DVB) matrices with sulfonic acid (SO^{3-}) groups as the functional groups.

☆ Cation-exchange resins are commonly used for the separation and purification of cations such as metal ions and proteins.

2.5.2. Anion-exchange resins

☆ Anion-exchange resins (anion exchangers) have positively charged functional groups that selectively retain negatively charged analytes (Table 1 and Figure 1).

☆ They are usually composed of polystyrene divinylbenzene (PS-DVB) matrices with quaternary ammonium ($N^+(CH_3)_3$) groups as the functional groups.

☆ Anion-exchange resins are employed for the separation and purification of anions such as organic acids, nucleic acids, and other negatively charged biomolecules.

Table 1: Table showing the charges on the cation- and anion-exchange resins.

Resin type	Cation-exchange resins (cation exchangers)	Anion-exchange resins (anion exchangers)
Charge of resin	–	+
Net charge on the molecule of interest	+	–
Running conditions	0.5-1.5 pH units below the pl of the molecule of interest	0.5-1.5 pH units above the pl of the molecule of interest

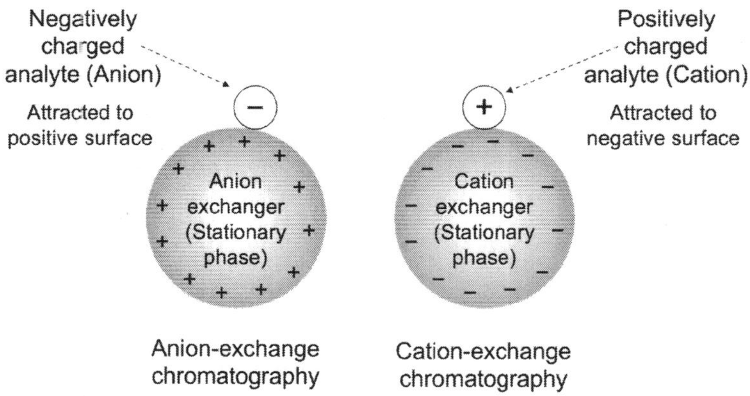

Figure 1: Schematic representation of a cation- and anion-exchange resin molecule showing the charges. Cation-exchange resin carries positive charges, while anion-exchange resin carries negative charges. Analyze ions with negative and positive charges are shown interacting with the resins.

3. Types of IEC

3.1. Cation-exchange chromatography

3.1.1. Principles

- ☆ Cation-exchange chromatography utilizes the reversible exchange of cations between the analyte molecules and a solid stationary phase containing negatively charged functional groups.
- ☆ The mechanism is based on electrostatic interactions between the positively charged analytes and the negatively charged functional groups on the resin.
- ☆ The principle behind cation-exchange chromatography is the selective retention of cations by the negatively charged functional groups (Figure 2). As the sample solution passes through the column packed with the cation-exchange resin, cations in the sample interact with the resin, binding to the functional groups. The extent of binding depends on the charge and concentration of the cations as well as the strength of the interaction between the cations and the functional groups.

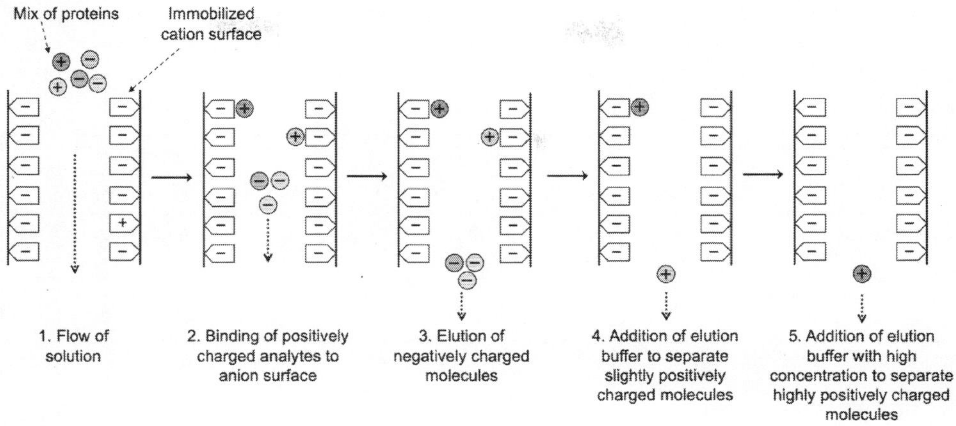

Figure 2: Step-by-step procedure of cation-exchange chromatography. (1) Sample containing both positively and negatively charged analytes is applied to the cation-exchange column. (2) Positively charged analytes are retained on the negatively charged resin. (3-5) Elution is performed using a mobile phase with increasing ionic strength or pH.

3.1.2. Types of cation-exchange resins

☆ There are different types of cation-exchange resins available for chromatographic applications (Table 2). These resins differ in chemical composition, structure, and properties. Some commonly used cation-exchange resins include:

1. **Sulfonated polystyrene-divinylbenzene (PS-DVB) resins:** These resins consist of a polystyrene matrix with sulfonic acid (SO^{3-}) groups as the functional groups. They offer high chemical stability and are commonly used in various industries for the separation and purification of cations.

2. **Carboxymethyl (CM) resins:** CM resins contain carboxymethyl groups ($-COO^-$) as the functional groups. These resins are suitable for the separation of alkaline earth metal cations and other metal ions.

3. **Polyacrylic acid (PAA) resins:** PAA resins have carboxylic acid groups (-COOH) as the functional groups. They are commonly used for the separation of alkali metal cations.

Table 2: Table showing resins commonly used in ion-exchange chromatography. The examples are a few representative types of resins. There are various other resins available with different functional groups, particle sizes, and pore sizes, with a range of options for specific separation needs.

Resin type	Functional group	Example
Strong cation exchange	Sulfonic acid (e.g., $-SO_3H$)	Sulfonated polystyrene resin
Weak cation exchange	Carboxylic acid (e.g., -COOH)	Carboxymethyl cellulose (CM) resin
Strong anion exchange	Quaternary ammonium (e.g., $(N^+(CH_3)_3)$)	Quaternary ammonium polystyrene resin
Weak anion exchange	Primary, secondary, tertiary amine	Diethylaminoethyl (DEAE) cellulose resin

Note

- ☆ The resins exhibit different ion-exchange properties and selectivities based on their functional groups.
- ☆ Strong cation exchange resins have a high affinity for cations and can be used to separate positively charged analytes. Weak cation exchange resins have a lower binding capacity for cations and are often used for more gentle separations.
- ☆ Strong anion exchange resins attract and retain anions, while weak anion exchange resins have lower binding capacity for anions.
- ☆ Examples of cation-exchange resins include Dowex 50W (strong acid cation-exchange resin based on sulfonated polystyrene), Amberlite IR-120 (strong acid cation-exchange resin with high exchange capacity), and Purolite C100 (macroporous, gel-type cation-exchange resin used for water softening applications).
- ☆ Examples of anion-exchange resins include Dowex 1X8 (strong base anion-exchange resin based on quaternary ammonium groups), Amberlite IRA-400 (type I strong base anion-exchange resin used for removing anions from water), and Purolite A400 (gel-type anion-exchange resin with high selectivity for various anions).

3.1.3. Applications

- ☆ Cation-exchange chromatography is commonly used in the pharmaceutical and biotechnology industries for the purification of proteins, peptides, and other biomolecules. It is particularly effective for separating proteins based on their charge differences.

- ☆ In environmental analysis, cation-exchange chromatography is used for the determination of heavy metal cations in water, soil, and other samples. It plays a crucial role in monitoring environmental pollutants and ensuring compliance with safety regulations.
- ☆ Cation-exchange chromatography is also suitable for the separation and analysis of metal ions in various industries, including mining, metallurgy, and wastewater treatment. It allows for the quantification and specialization of metal ions, aiding in quality control and process optimization.

3.1.4 Advantages and limitations

- ☆ Cation-exchange chromatography offers several advantages. It provides high selectivity for cations, allowing for efficient separation of complex mixtures. The technique is relatively easy to use and can be adapted to various scales, from analytical to preparative chromatography.
- ☆ Furthermore, it can be coupled with other chromatographic techniques, such as size exclusion chromatography or reversed-phase chromatography, to achieve comprehensive purification and characterization of complex samples.
- ☆ However, cation-exchange chromatography also has limitations. It may not be suitable for the separation of analytes with similar charges or closely related chemical properties. It requires careful optimization of the mobile phase composition and pH to achieve desired separations. In some cases, the resin may have limited stability or capacity for certain analytes, affecting the overall performance.

3.2. Anion-exchange chromatography

3.2.1. Principles

- ☆ Anion-exchange chromatography exploits the reversible exchange of anions between analyte molecules and a solid stationary phase containing positively charged functional groups.
- ☆ The mechanism is based on the electrostatic interactions between the negatively charged analytes and the positively charged functional groups on the resin.
- ☆ The principle behind anion-exchange chromatography is the selective retention of anions by the positively charged functional groups (Figure 3). As the sample solution passes through the column packed with the anion-exchange resin, anions in the sample interact with the resin, binding to the functional groups. The extent of binding depends on the charge and concentration of the anions as well as the strength of the interaction between the anions and the functional groups.

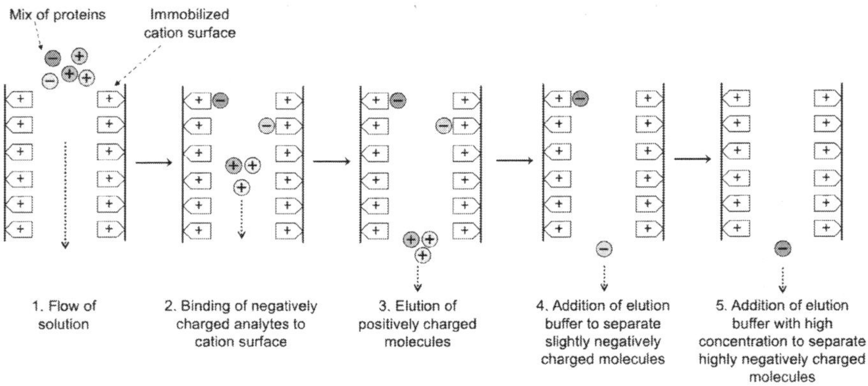

Figure 3: Step-by-step procedure of anion-exchange chromatography. (1) Sample containing both positively and negatively charged analytes is applied to the anion-exchange column. (2) Negatively charged analytes are retained on the positively charged resin. (3-5) Elution is performed using a mobile phase with increasing ionic strength or pH.

3.1.2. Types of anion-exchange resins

☆ There are different types of anion-exchange resins available for chromatographic applications (Table 2). These resins differ in chemical composition, structure, and properties. Some commonly used anion-exchange resins include:

1. **Quaternary ammonium (QA) resins:** QA resins consist of a polystyrene matrix with quaternary ammonium ($N^+(CH_3)_3$) groups as the functional groups. They offer high chemical stability and are commonly used in the purification and separation of anions.

2. **Tertiary amine (TA) resins:** TA resins have tertiary amine groups as the functional group. They are suitable for the separation of acidic compounds and are often used for the analysis of organic acids and other negatively charged analytes.

3. **Weak-base anion-exchange resins:** Weak-base resins contain weakly basic functional groups such as primary and secondary amines. They exhibit different selectivity and are used for the separation of specific anions under appropriate pH conditions.

3.1.3. Applications

☆ Anion-exchange chromatography is widely used in the pharmaceutical and biotechnology industries for the purification of biomolecules such as proteins, peptides, and nucleic acids. It is particularly effective for separating and purifying acidic proteins and other negatively charged molecules.

- ☆ In environmental analysis, anion-exchange chromatography is employed for the determination of anions, including inorganic anions such as nitrate, sulphate, and phosphate. It plays a crucial role in monitoring water quality, evaluating pollution levels, and ensuring compliance with regulatory standards.
- ☆ Anion-exchange chromatography is also used in the analysis of organic acids and small organic molecules in the food and beverage industry. It aids in the determination of organic acid profiles, flavour components, and contaminants.

3.1.4. Advantages and limitations

- ☆ Anion-exchange chromatography offers several advantages. It provides high selectivity for anions, allowing for efficient separation of complex mixtures. The technique is relatively easy to use and can be adapted to various scales, from analytical to preparative chromatography.
- ☆ Furthermore, it can be combined with other chromatographic techniques, such as size exclusion chromatography or reversed-phase chromatography, to achieve comprehensive purification and characterization of complex samples.
- ☆ However, anion-exchange chromatography has certain limitations. It may not be suitable for the separation of analytes with similar charges or closely related chemical properties. Optimization of the mobile phase composition and pH is crucial to achieving desired separations. The resin may have limited stability or capacity for certain analytes, impacting the overall performance.

4. Methodology

- ☆ Experimental design and methodology are crucial aspects of IEC.
- ☆ The experimental design involves determining the objectives of the study, selecting the appropriate sample preparation techniques, and choosing the suitable column, resin, buffer system, and elution strategy.
- ☆ The methodology encompasses the step-by-step procedures and techniques used in the experiment, including sample handling, column packing, buffer preparation, and elution. It outlines the specific details and parameters for each step to ensure reproducibility and reliability.
- ☆ By carefully designing the experiment and methodology, one can obtain accurate and meaningful results in IEC.

4.1. Sample preparation and handling

- ☆ In IEC, proper sample preparation and handling are crucial for obtaining accurate and reliable results.

- ☆ The sample should be properly prepared to ensure that the analytes of interest are in a suitable form for interaction with the ion-exchange resin. This may involve filtration, centrifugation, or dilution to remove particulate matter or adjust the sample concentration.

4.2. Column and resin selection

- ☆ The choice of column and resin depends on the specific requirements of the separation.
- ☆ Factors such as sample size, analyte properties, and desired resolution play a significant role in column and resin selection.
- ☆ Column size and resin bed height are determined based on the sample volume and desired separation efficiency.
- ☆ The resin should have the appropriate functional groups for the targeted analytes (cation or anion exchange). Additionally, resin particle size, porosity, and stability must be considered.

4.3. Buffer and pH optimization

- ☆ The selection and optimization of the buffer system and pH are critical in IEC.
- ☆ The buffer provides ionic strength and pH control to promote the desired ion-exchange interactions.
- ☆ The choice of buffer components and concentration should be based on the analyte properties and the resin's capacity.
- ☆ pH optimization is crucial as it affects the charge of the analytes and the resin, thereby influencing their interaction. The pH should be adjusted to maximize separation efficiency and resolution.

4.4. Elution strategies

- ☆ Elution strategies determine how analytes are eluted from the ion-exchange resin.
- ☆ Different elution strategies can be employed to achieve the desired separation. These include isocratic elution, where the eluent composition remains constant throughout the separation, and gradient elution, where the eluent composition is varied over time.
- ☆ Gradient elution allows for better resolution and separation of closely eluting peaks. The elution flow rate and volume should also be optimized to ensure sufficient separation and peak resolution.

4.5. Analysis and characterization

- ☆ After elution, the separated analytes need to be analyzed and characterized.
- ☆ This may involve various detection methods, such as UV-Vis spectroscopy, refractive index detection, or mass spectrometry.
- ☆ The choice of detection method depends on the analyte properties and the desired sensitivity and specificity.
- ☆ Quantification of analytes can be done by constructing calibration curves using known standards.
- ☆ The separated analytes can also be further characterized using other techniques, such as SDS-PAGE or capillary electrophoresis, to assess purity, molecular weight, or charge heterogeneity.

5. Factors Affecting IEC

- ☆ Several factors influence the performance of the IEC.
- ☆ Sample properties, such as analyte charge, concentration, and size, impact the interaction with the resin.
- ☆ Chromatography conditions, including buffer composition, pH, flow rate, and temperature, affect the ion-exchange interactions.
- ☆ Resin and column factors, such as resin type, particle size, porosity, and column dimensions, influence selectivity, resolution, and capacity.
- ☆ Careful consideration and optimization of these factors are essential for achieving optimal IEC performance and obtaining accurate and reliable results.

5.1. Sample properties

- ☆ The performance of IEC is influenced by various factors related to the properties of the sample being analyzed. Some key sample properties that can affect chromatography performance include:
  - ☆ **Analyte charge:** The charge of the analytes determines their interaction with the ion-exchange resin. IEC is highly selective based on charge, and analytes with higher charge densities tend to interact more strongly with the resin.
  - ☆ **Analyte concentration:** The concentration of the analytes affects their binding capacity on the resin. At high concentrations, the resin may become saturated, leading to reduced binding and lower separation efficiency. Conversely, at low concentrations, the analytes may exhibit weak interactions with the resin, resulting in poor separation.

- **Analyte size:** The size of the analytes can influence their diffusion rates within the resin matrix. Large analytes may have slower diffusion rates, leading to longer retention times and broader peaks.

5.2. Chromatography conditions

- The optimization of chromatography conditions is critical for achieving desired separation and performance. Factors related to chromatography conditions that can impact the performance of IEC include:
  - **Buffer composition and pH:** The choice of buffer components and pH is essential for controlling the ionic strength and optimizing the ion-exchange interactions. The buffer pH affects the analyte charge and the resin charge, thus influencing their interactions.
  - **Flow rate:** The flow rate of the mobile phase affects the residence time of the analytes on the resin. A too high flow rate may result in reduced resolution, while a too low flow rate may cause longer run times.
  - **Temperature:** The temperature can impact the kinetics of the ion-exchange interactions and the analyte diffusion within the resin. Changes in temperature can affect the selectivity and resolution of the separation.

5.3. Resin and column factors

- The selection and properties of the resin and column play a significant role in IEC performance. Factors related to the resin and column that can influence performance include:
  - **Resin type:** The type of ion-exchange resin used, such as strong or weak ion-exchangers, affects the selectivity and binding capacity for specific analytes. Different resins have varying affinities for different types of ions.
  - **Resin particle size and porosity:** The particle size and porosity of the resin can impact the column efficiency and resolution. Smaller resin particles and higher porosity can provide better separation and higher resolution.
  - **Column length and diameter:** The column length and diameter affect the separation efficiency and capacity of the column. Wider columns generally offer higher resolution.

6. Techniques and Optimization Strategies

6.1. Sample preparation and handling

- Sample preparation and handling are crucial steps in optimizing IEC. Proper sample preparation ensures that the analytes are in a suitable form for interaction with the ion-exchange resin.

- ☆ Several techniques can be employed during sample preparation, including filtration, centrifugation, and dilution. These techniques help remove particulate matter, cellular debris, or high-concentration impurities that may interfere with the chromatographic process.
- ☆ Careful handling of the sample is also important to minimize the risk of sample degradation or contamination.

6.2. Resin and column optimization

- ☆ The selection and optimization of the resin and column are essential for achieving optimal IEC results. The resin should have the appropriate functional groups for the targeted analytes.
- ☆ Factors to consider during resin and column optimization include resin type, particle size, and porosity, as well as column dimensions.
- ☆ Different resins may have different selectivities and binding capacities, so it is crucial to select the most suitable resin for the analytes of interest.
- ☆ Additionally, the resin particle size and porosity affect the separation efficiency, resolution, and pressure drop in the column.
- ☆ Optimizing the column dimensions based on the sample volume and desired separation efficiency is crucial for achieving the best results.

6.3. Buffer and pH optimization

- ☆ Buffer and pH optimization are critical for achieving optimal ion-exchange interactions. The buffer provides the necessary ionic strength and pH control to promote the desired ion-exchange interactions between the analytes and the resin.
- ☆ The choice of buffer components and concentrations should be based on the analyte properties and the resin's capacity.
- ☆ The pH of the buffer plays a significant role in the charge of the analytes and the resin, influencing their interactions.
- ☆ Optimization of the buffer and pH is essential for maximizing separation efficiency and resolution.
- ☆ Additionally, the ionic strength of the buffer should be optimized to enhance the binding capacity and selectivity of the resin.

6.4. Elution strategies

- ☆ Elution is the process of removing bound analytes from the resin after the separation has taken place.
- ☆ Elution strategies play a critical role in ion-exchange chromatography, allowing for the selective and efficient elution of analytes from the ion-exchange resin.

☆ The selection of the appropriate elution strategy depends on the nature of the analytes, the desired separation goals, and the specific chromatographic system.

☆ By carefully designing and optimizing the elution strategy, one can achieve efficient and selective elution of analytes in IEC, leading to improved resolution, separation, and quantification of compounds.

☆ Several elution strategies can be employed based on the specific requirements of the analysis.

1. **Isocratic elution:** In isocratic elution, a constant composition eluent is used throughout the separation process. It is suitable when the analytes have similar binding strengths to the resin and need to be eluted together. Isocratic elution is relatively simple, but it may not provide optimal resolution for complex mixtures with closely eluting peaks.
2. **Gradient elution:** Gradient elution involves changing the eluent composition over time. It is useful when the analytes have different binding strengths to the resin and need to be separated based on their affinity. By gradually increasing the eluent strength, analytes with weaker binding affinity are eluted first, followed by those with stronger binding. It provides better resolution and separation for complex samples with overlapping peaks.
3. **Step elution:** Step elution involves abruptly changing the eluent composition or concentration to selectively elute specific analytes. It is used when the analytes have significantly different binding strengths. By adjusting the eluent conditions, analytes with weaker or stronger binding affinity can be selectively released from the resin. Step elution can enhance selectivity and improve resolution in certain applications.
4. **pH gradient elution:** pH gradient elution utilizes a varying pH to selectively elute analytes based on their p*K*a values. By changing the pH of the eluent, analytes with different acid-base properties can be selectively released from the resin. It is particularly effective for separating analytes that have different ionization behaviours at different pH values.

Note

☆ NaCl is the most common eluent in IEC. It is used in isocratic, gradient, and step elution methods..

- ☆ In isocratic elution, a constant concentration of NaCl is used as the eluent throughout the chromatography process. This strategy is effective when the analytes of interest have similar binding affinities to the ion-exchange resin. Isocratic elution with NaCl provides a simple and consistent elution condition, allowing for the simultaneous elution of all analytes without significant changes in the eluent strength.
- ☆ In gradient elution, the concentration of NaCl is varied over time. This strategy is particularly useful when separating analytes with different binding affinities to the resin. By gradually increasing the NaCl concentration, analytes with weaker binding affinities are eluted first, followed by those with stronger binding affinities. Gradient elution with NaCl enables improved resolution and separation of closely eluting peaks, enhancing the selectivity and efficiency of the chromatographic process.
- ☆ In step-wise elution, the eluent conditions are abruptly changed to selectively elute specific analytes using NaCl. This strategy is employed when analytes have significantly different binding affinities. By adjusting the NaCl concentration or pH, analytes with weaker or stronger binding affinities can be selectively released from the resin, enhancing selectivity and improving resolution in specific applications.

7. Applications of IEC

- ☆ IEC is a versatile separation technique that finds applications in various fields of research, industry, and healthcare.
- ☆ It is widely used for the separation, purification, and analysis of charged molecules based on their interactions with ion-exchange resins. Below are a few important applications of IEC:

7.1. Protein purification

- ☆ IEC is extensively employed for the purification of proteins and peptides.
- ☆ It can separate proteins based on their net charge and ionic interactions with the resin.
- ☆ Strong and weak ion-exchange resins are used to selectively bind and elute proteins of interest, facilitating their purification from complex mixtures.

7.2. Pharmaceutical analysis

- ☆ IEC plays a crucial role in pharmaceutical analysis. It enables the separation and quantification of ionic drug substances, counterions, and impurities in pharmaceutical formulations.

- ☆ IEC methods are employed in quality control laboratories to ensure the safety, efficacy, and purity of pharmaceutical products.

7.3. Environmental analysis

- ☆ IEC is utilized for the analysis of environmental samples to determine the concentration of inorganic and organic ions.
- ☆ It aids in monitoring water quality, identifying pollutants, and assessing environmental contamination. It enables the separation and quantification of ions such as heavy metals, nutrients, and organic acids in environmental samples.

7.4. Food and beverage analysis

- ☆ IEC is employed in the food and beverage industry for quality control and analysis.
- ☆ It enables the determination of various ions, including organic acids, amino acids, vitamins, and minerals, in food and beverage products.
- ☆ IEC is essential for assessing product quality, identifying adulterants, and ensuring regulatory compliance.

7.5. Biopharmaceutical development

- ☆ IEC is widely used in biopharmaceutical development for the analysis, purification, and characterization of biologics such as monoclonal antibodies, vaccines, and recombinant proteins.
- ☆ It aids in the separation of closely related variants, determination of charge heterogeneity, and assessment of protein quality attributes.

7.6. Purification of nucleic acids

- ☆ IEC is employed for the purification of nucleic acids, including DNA and RNA.
- ☆ It can separate and purify nucleic acids based on their charge and ionic interactions with the resin.
- ☆ It is a valuable tool in molecular biology research, genetic analysis, and forensic sciences.

7.7. Industrial processes

- ☆ IEC finds applications in various industrial processes, including water treatment, metal recovery, and pharmaceutical manufacturing.
- ☆ It helps in the removal of unwanted ions, separation of valuable metals, and purification of industrial products.

8. Advancements and Innovations in IEC

☆ Advancements and innovations in IEC have significantly improved its performance, efficiency, and applicability across various scientific and industrial fields. Below are some notable advancements of IEC.

1. **Novel ion-exchange resins:** The development of new ion-exchange resins with improved selectivity, higher capacity, and enhanced stability has expanded the capabilities of IEC. These resins exhibit better chemical and mechanical stability, enabling the separation of a wide range of analytes with high resolution and efficiency.
2. **Multi-modal chromatography:** Multi-modal ion-exchange resins, which possess multiple functional groups on a single resin, have emerged as a powerful tool in separation science. By combining different interaction mechanisms, such as ion-exchange, hydrophobic, and affinity interactions, multi-modal chromatography provides enhanced selectivity and allows for complex separations.
3. **High-throughput and automated systems:** Automation has revolutionized IEC by increasing sample throughput, improving reproducibility, and reducing human error. High-throughput systems enable the simultaneous processing of multiple samples, while automated systems offer precise control over parameters, such as sample loading, elution, and column regeneration, ensuring consistent and efficient separations.
4. **Integration with other techniques:** IEC has been successfully integrated with other chromatographic methods, such as size exclusion chromatography (SEC) and reverse-phase chromatography (RPC), as well as mass spectrometry (MS). These integrations enable multidimensional separations, improved selectivity, and enhanced detection capabilities, leading to comprehensive analysis of complex samples.
5. **Miniaturization and microfluidic systems:** Miniaturization of IEC has led to the development of microfluidic systems, which offer advantages such as reduced sample and reagent consumption, faster analysis times, and increased sensitivity. Microfluidic IEC holds promise for point-of-care diagnostics, environmental monitoring, and high-throughput analysis.

06 Gel Filtration Chromatography

CHAPTER SUMMARY

- Gel filtration chromatography (GFC) is a powerful technique used for separating and analysing molecules based on their size and molecular weight.
- It separates molecules based on their size using a porous stationary phase that allows smaller molecules to enter the pores and elute later, while larger molecules pass through more quickly.
- In GFC, a sample is loaded onto a column packed with a porous gel matrix. As the sample flows through the column, molecules of different sizes are separated based on their ability to enter the pores of the gel. Smaller molecules get trapped, while larger ones elute faster. The eluted fractions are collected and analysed to determine the molecular weight distribution of the sample.
- Factors that affect GFC include the size and shape of the molecules, column characteristics, buffer composition, flow rate, and sample concentration.
- GFC utilizes various types of matrix materials, such as agarose and crosslinked dextran, which form porous beads that selectively separate molecules based on their size and shape.
- GFC is commonly used for protein purification, determination of molecular weight, protein complex analysis, and separation of biomolecules based on size.

1. Introduction

- ☆ Gel filtration chromatography (GFC) is a widely used separation technique in biochemistry and biotechnology.
- ☆ It is also known as gel permeation chromatography (GPC) or size exclusion chromatography (SEC), or molecular sieve chromatography (MSC).
- ☆ GFC separates molecules based on their size or molecular weight. The principle behind GFC is that molecules of different sizes will penetrate different depths into a porous gel matrix, thus leading to their separation.

- GFC originated in the late 1950s when scientists discovered that certain porous resins could be used to separate molecules based on their size. Since then, numerous gel matrices have been developed, and GFC has become one of the most commonly used methods for separating and purifying biomolecules.
- The basic components of GFC include a column filled with a porous matrix, a sample containing the molecules to be separated, and a mobile phase that carries the sample through the column. The porous matrix is typically made of cross-linked dextran or agarose, and the pore size is chosen based on the size of the molecules to be separated.
- When the sample is loaded onto the column, the larger molecules are excluded from entering the porous matrix and thus travel through the column faster than the smaller molecules. On the other hand, the smaller molecules penetrate deeper into the matrix and thus take longer to travel through the column. As a result, the molecules are separated based on their size, with the larger molecules eluting from the column first and the smaller molecules eluting later.
- GFC has a wide range of applications in biochemistry and biotechnology. It is commonly used for protein purification, enzyme isolation, and the separation of nucleic acids and carbohydrates. It is also used to analyse macromolecules, such as in determining the molecular weight of proteins.

2. Principles

- GFC is a powerful technique used for separating and purifying biomolecules based on their size or molecular weight.
- The basic principle of GFC involves the use of a porous gel matrix to separate molecules according to their size.

2.1. Molecular size separation

- The primary principle of GFC is the separation of molecules based on their size. Larger molecules cannot penetrate the pores of the gel matrix and therefore elute first, while smaller molecules penetrate more deeply and take longer to elute from the column (Figure 1). This results in the separation of molecules based on their size or molecular weight.
- The separation of molecules based on their size is achieved by selecting a gel matrix with a pore size that is appropriate for the size of the molecules being separated. For example, a larger pore size will allow larger molecules to penetrate the matrix, while a smaller pore size will restrict the penetration of larger molecules.

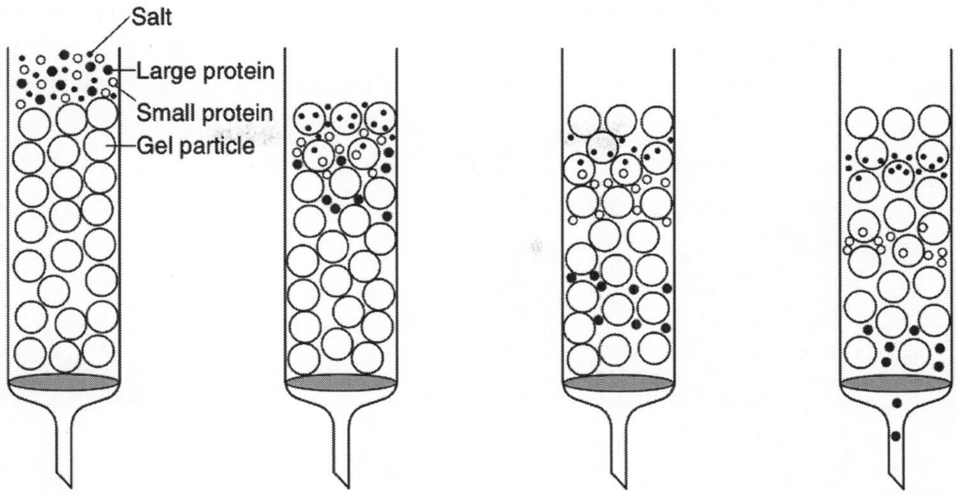

Figure 1: Procedure of gel filtration chromatography. A sample mixture containing different-sized molecules is loaded onto the column, and molecules are separated based on their size as they pass through the matrix. Eluted fractions are collected and analysed.

2.2. Separation mechanisms

☆ The separation of molecules in the GFC occurs due to two mechanisms: steric exclusion and flow rate.

1. **Steric exclusion** is the primary mechanism responsible for separating molecules based on their size. Steric exclusion occurs due to the steric hindrance caused by the size of the molecule. A large molecule will not be able to penetrate the pores of the gel matrix and will therefore elute first. This is because the pore size of the matrix is smaller than the size of the molecule, and the molecule cannot fit into the pores.
2. **Flow rate** is another mechanism that contributes to the separation of molecules in the GFC. The flow rate of the mobile phase affects the rate at which molecules elute from the column. Smaller molecules can travel through the matrix more easily and elute faster, while larger molecules are slowed down due to their size, resulting in their separation from smaller molecules.

2.3. Characteristics of matrices

☆ The porous matrix used in GFC is typically made of cross-linked dextran or agarose.

- ☆ The pore size of the matrix is selected based on the size of the molecules to be separated. A larger pore size will allow larger molecules to penetrate the matrix, while a smaller pore size will restrict the penetration of larger molecules.
- ☆ The matrix used in GFC should be stable, inert, and compatible with the molecules being separated.
- ☆ Different matrices have different characteristics, such as binding capacity, resolution, and column stability, and the choice of a matrix depends on the specific needs of the separation process. For example, some matrices may have a higher binding capacity for certain molecules, while others may have a higher resolution and be better suited to separate closely related molecules.

Note

- ☆ GFC separates molecules based on their size, but it can also be influenced by other interactions between the molecules and the gel matrix. These interactions include hydrophobic interactions, electrostatic interactions, and hydrogen bonding. In some cases, these interactions can cause molecules to bind to the gel matrix, which can affect the separation process.
- ☆ Therefore, it is important to select a gel matrix that is suitable for the specific molecule being separated to minimize the effects of these interactions.

3. Methodology

- ☆ The procedure of GFC involves the following steps: sample preparation, column preparation, sample loading, and elution.
- ☆ Understanding the procedure of GFC can aid in the selection of appropriate matrices and the optimization of separation protocols for specific biomolecules.

1. **Sample preparation:** The first step in GFC is the preparation of the sample. The sample must be dissolved in a buffer that is compatible with the gel matrix and the molecules being separated. The buffer should be chosen based on the pH and ionic strength required for optimal separation. The sample is typically centrifuged or filtered through a 0.2-micron filter to remove any particulate matter that may clog the column. The sample concentration must also be determined to ensure that the column is not overloaded.

2. **Column preparation:** The next step is the preparation of the gel filtration column. The column is usually composed of a glass or plastic tube that is packed with a porous gel matrix. The gel matrix is chosen based on the size of the molecules being separated. Sephadex, Sephacryl, and BioGel are some common gel matrices used in GFC. The column is packed in a column holder and equilibrated with a buffer solution to ensure that the matrix is hydrated and to remove any air bubbles.
3. **Sample loading:** Once the column is prepared, the sample is loaded onto the top of the column. The sample is applied in a small volume to prevent overloading the column and to ensure that the sample is evenly distributed throughout the gel matrix. The sample is allowed to enter the column under gravity or using a pump system. As the sample enters the column, the larger molecules are excluded from the gel matrix and pass through the column more quickly than the smaller molecules.
4. **Elution:** The final step in GFC is the elution of the separated molecules from the column. Elution is typically performed using a buffer solution that is compatible with the molecules being separated. As the buffer solution flows through the column, the molecules are separated based on their size and elute from the column at different times. The eluted fractions are collected and analysed to determine the presence and purity of the separated molecules.

☆ After elution, the fractions are analysed to determine the purity and concentration of the separated molecules. SDS-PAGE and UV spectroscopy are some common techniques used for analysing eluted fractions. The fractions can also be concentrated or desalted using techniques such as ultrafiltration or dialysis.

Note

☆ After the completion of the GFC run, the column must be cleaned and stored properly. Column cleaning and storage are essential steps in maintaining the functionality and accuracy of GFC columns.

☆ Impurities from the sample can adhere to the gel matrix during separation, reducing the efficiency of the column and causing retention time shifts. Proper cleaning with suitable agents and storage in an appropriate environment can prevent the buildup of contaminants and ensure the longevity of GFC columns. Recommended cleaning methods include water, buffer, detergent, and high salt solutions, while buffer solution, ethanol, and dry storage are recommended for column storage.

☆ Regular column cleaning and proper storage can ensure optimal performance and accurate separation for an extended period.

4. Factors Affecting GFC

- ☆ Several factors can affect the performance of GFC, including pH and salt concentration, flow rate, column length and diameter, sample size, and matrix characteristics.
- ☆ These factors should be optimized based on the sample properties and the desired separation results.
- ☆ Proper optimization of these factors can lead to accurate separation and efficient purification of biomolecules.

4.1. pH and salt concentration

- ☆ The pH and salt concentration of the buffer used for GFC can significantly affect the separation of biomolecules.
- ☆ The pH of the buffer affects the ionization of functional groups on the gel matrix and the sample molecules, which can affect their interaction and separation.
- ☆ Higher salt concentrations can lead to a decrease in the separation efficiency of the column due to ionic screening, reducing the interactions between the sample and the gel matrix.
- ☆ Therefore, it is essential to optimize the pH and salt concentration of the buffer to achieve optimal separation.

4.2. Flow rate

- ☆ The flow rate of the buffer through the GFC column can affect the separation efficiency and resolution.
- ☆ A high flow rate can reduce the interaction time between the sample and the gel matrix, leading to lower resolution and reduced separation efficiency.
- ☆ A low flow rate can lead to excessive band broadening and longer separation times.
- ☆ Therefore, the flow rate should be optimized to achieve the best separation results.

4.3. Column length and diameter

- ☆ The length and diameter of the GFC column can significantly affect the separation performance.
- ☆ Longer columns generally provide better resolution and separation efficiency but at the expense of longer separation times.
- ☆ Wider columns can handle larger sample volumes but at the expense of lower resolution.

- ☆ Therefore, the length and diameter of the column should be optimized based on the sample size and desired resolution.

4.4. Sample size

- ☆ The sample size used for GFC can also affect the separation performance.
- ☆ Overloading the column with a large sample can lead to band broadening, peak distortion, and reduced separation efficiency.
- ☆ On the other hand, using too small a sample can lead to poor resolution and a low signal-to-noise ratio.
- ☆ Therefore, it is important to use an appropriate sample size for the desired separation.

4.5. Matrix characteristics

- ☆ The matrix, also known as the stationary phase, is a key component of GFC and determines the separation performance. The choice of a matrix is critical to achieving optimal separation of the target molecule.
- ☆ The pore size, particle size, and functional groups on the gel matrix can all affect the interactions between the sample and the matrix.
- ☆ A smaller pore size can lead to better resolution and separation efficiency for smaller molecules, while a larger pore size can accommodate larger molecules.
- ☆ The particle size can also affect the flow rate and resolution. Smaller particles generally provide better resolution but can lead to longer separation times.
- ☆ The functional groups on the gel matrix can also affect the separation, as they can interact with the functional groups on the sample molecules.
- ☆ Therefore, the choice of gel matrix should be optimized based on the sample properties and desired separation.

5. Types of Matrices in GFC

- ☆ The choice of a matrix in GFC is a crucial factor in achieving optimal separation of molecules based on their size.
- ☆ Agarose and dextran-based matrices are the most commonly used matrices in GFC due to their high mechanical stability, biocompatibility, and low non-specific binding. A representative schematic of a section through a Superdex particle is shown in Figure 2.
- ☆ Polyacrylamide, polystyrene, and silica-based matrices are also available and are suitable for separating small molecules, peptides, and proteins.
- ☆ The choice of a matrix depends on the specific application and the size of the molecule(s) being separated.

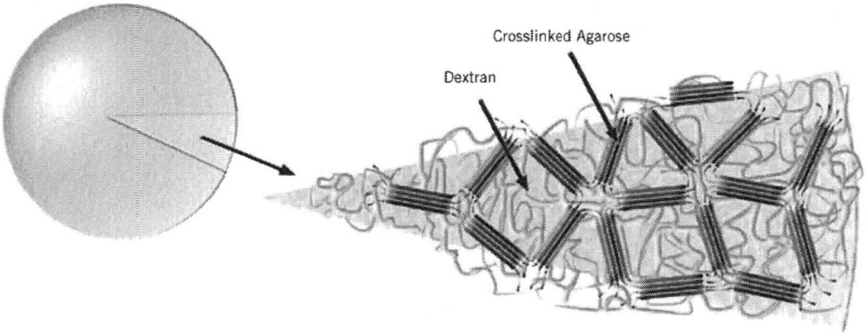

Figure 2: Representative schematic of a section through a Superdex matrix particle. In Superdex, the dextran chains are covalently linked to a highly cross-linked agarose matrix.

5.1. Agarose-based matrices

- Agarose is a linear polysaccharide composed of agarobiose units.
- Agarose-based matrices are widely used in GFC due to their high mechanical stability, biocompatibility, and low non-specific binding.
- These matrices are available in different pore sizes, and their separation performance can be tuned by adjusting the pore size.
- **Sepharose, Superdex,** and **Sephacryl** are agarose-based matrices used in GFC. They are made from cross-linked agarose and are available in different pore sizes.
- These matrices are used to purify proteins, nucleic acids, and other biomolecules.

5.2. Dextran-based matrices

- Dextran is a linear polysaccharide composed of glucose units.
- Dextran-based matrices are also widely used in GFC due to their high mechanical stability, biocompatibility, and low non-specific binding.
- These matrices are available in different pore sizes, and their separation performance can be tuned by adjusting the pore size.
- **Sephadex** is a dextran-based matrix used in GFC. It is made from cross-linked dextran and is available in different pore sizes. Sephadex is widely used to purify proteins, nucleic acids, and enzymes.

5.3. Polyacrylamide-based matrices

- Polyacrylamide-based matrices are made by cross-linking polyacrylamide with a cross-linking agent.

☆ These matrices are available in different pore sizes and are widely used to separate small molecules, peptides, and proteins.

☆ **Bio-Gel** is a polyacrylamide-based matrix used in GFC. It is made from cross-linked polyacrylamide and is available in different pore sizes. Bio-Gel is widely used to purify proteins, nucleic acids, and enzymes.

5.4. Polystyrene-based matrices

☆ Polystyrene-based matrices are made by cross-linking polystyrene with a cross-linking agent.

☆ These matrices are available in different pore sizes and are widely used to separate small molecules, peptides, and proteins.

☆ **Toyopearl** is a polystyrene-based matrix used in GFC. It is made of highly cross-linked and rigid polystyrene beads. It is used to purify proteins, nucleic acids, and enzymes.

6. Important Parameters in GFC

☆ The process of separation by GFC can be characterized by several parameters that describe the properties of the matrix, the molecular size, and the separation efficiency.

6.1. Retention time (tR)

☆ The most basic equation in GFC is the retention time (tR) of a molecule, which is defined as the time it takes for the molecule to pass through the column.

☆ The retention time is influenced by the size of the molecule, the pore size of the matrix, and the flow rate of the mobile phase.

☆ The retention time can be used to calculate the molecular weight of the molecule using the calibration curve equation, which relates the retention time to the molecular weight of a set of standard molecules.

6.2. Retention volume (V_r)

☆ Retention volume (V_r) is the volume of the mobile phase required for a molecule to elute from the column. It can be calculated by the equation:

$$V_r = V_e - V_0$$

where V_r is the retention volume of the analyte, V_e is the elution volume, and V_0 is the void volume.

☆ The retention volume can be used to calculate the molecular weight of the analyte using the following equations:

$$MW = K \times V_r$$

where MW is the molecular weight, K is a calibration constant specific to the gel filtration column, and V_r is the retention volume.

☆ The molecular weight can also be calculated using the following equation:

$$K_{av} = \frac{(V_e - V_0)}{(V_t - V_0)}$$

where K_{av} is the partition coefficient or the distribution coefficient of the molecule, and V_t and V_0 are the total and void volumes, respectively.

☆ By comparing the Kav values of unknown molecules with those of known standards of different molecular weights, the molecular weight of the unknown molecules can be estimated.

☆ A diagrammatic representation of V_0 and V_t is shown in Figure 3.

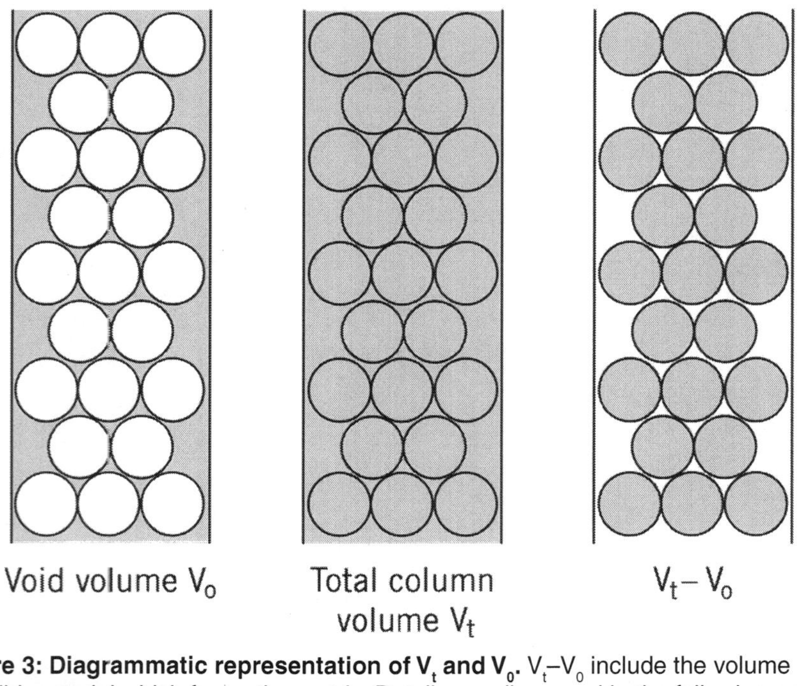

Figure 3: Diagrammatic representation of V_t and V_0. V_t–V_0 include the volume of the solid material which forms the matrix. Details are discussed in the following note.

Note

☆ **V_0 (void volume)** is the volume of the mobile phase that passes through the column without interacting with the stationary phase. It is the volume of the interstitial spaces between the beads in the column.

☆ **V_t (total volume)** is the volume of the mobile phase required to elute all the molecules from the column.

- ☆ **V_e (elution volume)** is the volume of the mobile phase at which a particular molecule is eluted from the column. It is the volume of the mobile phase that interacts with the stationary phase and causes the molecule to be separated from the other molecules.
- ☆ **V_0-V_t (dead volume)** is the volume of the solvent that has entered the stationary phase and is represented by the solvent front. It is the difference between the void volume (V_0) and the total volume (V_t) of the column.

6.3. Resolution (R)

- ☆ The resolution (R) of GFC describes the degree of separation between two adjacent peaks in the chromatogram.
- ☆ Resolution is calculated by dividing the difference between the retention times of two adjacent peaks by the sum of their peak widths at half height.
- ☆ The resolution is affected by several factors, including the column length and diameter, the pore size and distribution of the matrix, the mobile phase composition, and the flow rate.

6.4. Void volume (V_0)

- ☆ The void volume (V_0) represents the total volume of the spaces between the beads of the gel matrix that are not accessible to the sample molecules. It can be calculated by the equation:

$$V_0 = V_e - V_t$$

where V_e is the elution volume of the void marker, and V_t is the total bed volume of the column.

- ☆ The void marker used is a molecule that is too large to enter the pores of the matrix; thus, it elutes in the void volume. The void volume is calculated by injecting a small molecule, such as blue dextran, into the column and measuring its retention time.
- ☆ The total volume of the column is then subtracted from the retention volume of the blue dextran to obtain the void volume.
- ☆ The void volume is used to calculate the bed volume (V_t) of the column, which represents the volume of the matrix and the sample molecules.
- ☆ A theoretical chromatogram is shown in Figure 4.

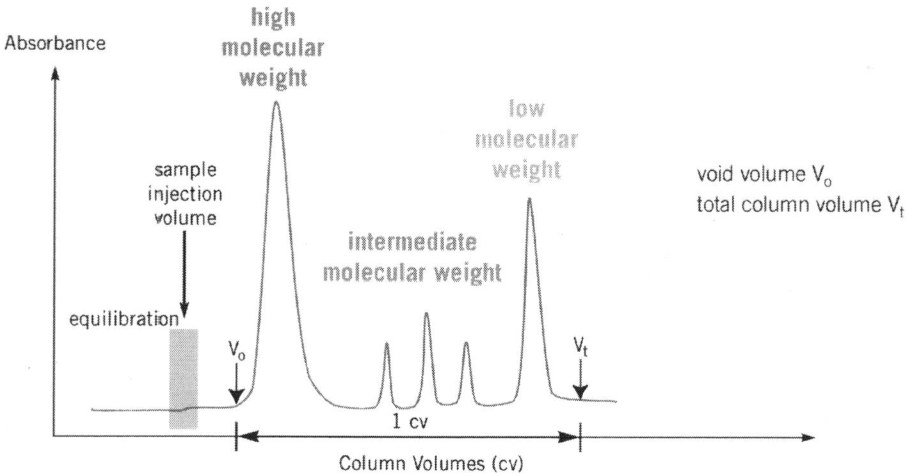

Figure 4: A theoretical chromatogram. The terms are described in the main text.

6.5. Plate number (N)

- The separation efficiency of GFC can be evaluated using the plate number (N), which represents the number of theoretical plates per unit length of the column.
- The plate number (N) is a measure of column efficiency in GFC, which represents the number of theoretical plates per unit length of the column. It can be calculated by the equation:

$$N = 5.54\left(\frac{V_r}{W_{1/2}}\right)^2 \text{ or } N = 16\left(\frac{tR}{W}\right)^2$$

where Vr is the retention volume of a solute, tR is retention time, and W is the width of the solute peak at half height.

- In practice, the plate number can be estimated by measuring the width of a solute peak at half height and the retention time.
- The plate number is influenced by several factors, including the column length and diameter, the flow rate, the packing density and uniformity of the matrix, and the sample concentration.

6.6. Peak height (H)

- Plate height (H) is another parameter that is used to measure column efficiency, which is the length of a column that is equivalent to one theoretical plate. It can be calculated using the equation:

$$H = \frac{L}{N}$$

where L is the column length, and N is the plate number.

- A higher plate number and a lower plate height indicate better column efficiency and higher resolution in chromatography.

6.7. Distribution coefficient (K_d)

- The distribution coefficient (K_d) is defined as the ratio of the concentration of a solute in the stationary phase (matrix phase) to the concentration of the solute in the mobile phase at *equilibrium.* '
- It is calculated by dividing the concentration of the molecule in the matrix phase by the concentration in the mobile phase at equilibrium.

$$K_d = \frac{[C_s]}{[C_m]}$$

where $[C_s]$ is the equilibrium concentration of the solute in the stationary phase and $[C_m]$ is the equilibrium concentration of the solute in the mobile phase.

- It is a measure of the affinity of a molecule for the matrix.
- It is used to predict the elution behaviour of a solute in GFC.
- The distribution coefficient is influenced by several factors, such as pH, salt concentration, size and charge of the solute, size of the molecule, as well as the properties of the matrix.

6.8. Partition coefficient (K) and capacity factor (k')

- The partition coefficient (K) is defined as the ratio of the concentration of a solute in the stationary phase to the concentration of the solute in the mobile phase at a *specific point* in the elution profile.
- It is also calculated using the equation:

$$K = \frac{[C_s]}{[C_m]}$$

- The capacity factor, k', is defined as the ratio of the retention volume of a solute to the void volume of the column. It can be calculated using the equation:

$$k' = \frac{V_r - V_0}{V_0}$$

where V_r is the retention volume of the solute, and V_0 is the void volume of the column.

- ☆ The capacity factor is a measure of the extent to which a solute interacts with the stationary phase. It is useful for optimizing the separation conditions in GFC.

Note

- ☆ In GFC, the distribution coefficient (K_d) and partition coefficient (K) are two terms used to describe the binding affinity of a solute to a stationary phase.
- ☆ The key difference between the two is the nature of the stationary phase. K_d is used to describe the binding affinity of a solute to the porous matrix of the gel filtration column, while K is used to describe the binding affinity of a solute to a non-porous stationary phase, such as in reverse-phase chromatography.
- ☆ K_d is calculated at equilibrium, while K is calculated at a specific point in the elution profile.

7. Applications

- ☆ GFC is a widely used technique for the separation and purification of biomolecules based on their size.
- ☆ Due to its simplicity, versatility, and effectiveness, GFC has found many applications in various fields, including biochemistry, biotechnology, pharmaceuticals, and food science.
- ☆ GFC can be used to purify proteins, enzymes, nucleic acids, viruses, polymers, and lipids.
- ☆ Proper optimization of GFC conditions can lead to accurate separation and efficient purification of biomolecules.

7.1. Protein purification

- ☆ GFC is widely used to purify proteins from complex mixtures such as cell lysates or culture supernatants.
- ☆ The technique can be used for both small-scale and large-scale purification of proteins, depending on the column size and sample volume.
- ☆ GFC can effectively separate proteins based on their size and shape, enabling the isolation of pure proteins from complex mixtures.

7.2. Enzyme purification

- ☆ GFC is an effective technique to purify enzymes, as it can separate them based on their size and shape.
- ☆ GFC can be used to purify enzymes from crude extracts or fermentation broths, enabling the isolation of pure enzymes with high activity.

7.3. Nucleic acid purification

- ☆ GFC can also be used to purify nucleic acids, such as DNA and RNA.
- ☆ The technique can effectively separate nucleic acids based on their size, enabling the isolation of pure nucleic acids from complex mixtures.
- ☆ GFC is often used in combination with other techniques, such as column chromatography or electrophoresis, for the isolation of high-quality nucleic acids.

7.4. Virus purification

- ☆ GFC is an effective technique to purify viruses from complex mixtures.
- ☆ The technique can be used to separate virus particles based on their size and shape, enabling the isolation of pure virus particles.
- ☆ GFC is often used in combination with other techniques, such as ultracentrifugation or ion exchange chromatography, to isolate high-quality virus particles.

7.5. Fractionation of polymers

- ☆ GFC can be used to fractionate polymers based on their size and molecular weight.
- ☆ The technique can be used to separate different polymer fractions, enabling the isolation of pure polymers with well-defined properties.
- ☆ GFC is often used in polymer chemistry to purify and characterize polymers.

7.6. Separation of lipids

- ☆ GFC can also be used to separate lipids based on their size and shape.
- ☆ The technique can be used to separate different lipid fractions, enabling the isolation of pure lipids with well-defined properties.
- ☆ GFC is often used in food science to purify and characterize lipids.

7.7. Desalting

- ☆ Desalting is a commonly used technique in biochemistry to remove salts and other low-molecular-weight solutes from biomolecules, such as proteins or nucleic acids.
- ☆ GFC is a powerful tool for desalting, as it separates molecules based on their size, allowing small solutes such as salts to pass through the gel matrix and be eluted first, while larger biomolecules like proteins are retained longer in the column.
- ☆ The desalting process can be optimized by choosing the appropriate column size and optimal flow rate to ensure efficient separation.

☆ Desalting is vital in protein purification, as high salt concentration can affect protein stability and activity.

☆ It is also used to remove unwanted small molecules or buffer components that interfere with downstream applications such as enzyme assays or structural studies.

07 Affinity Chromatography

CHAPTER SUMMARY

- Affinity chromatography is a powerful separation method that exploits specific interactions between a target molecule and an immobilized ligand, enabling purification of biomolecules.
- The principle is based on the specific binding affinity between a target molecule and a ligand immobilized on a solid support, allowing for selective separation and purification of the target molecule from a complex mixture.
- The types of matrix include agarose, cellulose, sepharose, and magnetic beads, which serve as the solid support for immobilizing the ligands and facilitating the selective binding of target molecules.
- In affinity chromatography, a specific ligand is immobilized onto a solid support matrix. The sample containing the target molecules is then applied, and the bound molecules are selectively eluted using a competing ligand or change in buffer conditions.
- Affinity chromatography finds applications in various fields, including protein purification, antibody purification, enzyme isolation, drug discovery, and biomolecule characterization.

1. Introduction

- Affinity chromatography is a widely used method in biochemistry, molecular biology, and biotechnology to isolate and purify specific biomolecules from complex mixtures.
- The term "Affinity" signifies the degree to which a molecule associates with another molecule.
- The method is based on the selective binding between a biomolecule of interest and a specific ligand that is immobilized on a solid support (such as agarose or sepharose).
- Affinity chromatography was first introduced as a method for purifying enzymes in the 1960s. It was initially used to separate enzymes based on their ability to bind to a specific substrate or inhibitor. Since then,

the method has been refined and expanded to include a wide range of biomolecules, including proteins, nucleic acids, carbohydrates, and small molecules.

- ☆ Affinity chromatography has become an essential tool in the field of protein purification, as it allows for the isolation of a specific protein from a complex mixture of proteins with high purity and yield.
- ☆ It is widely used in the production of biopharmaceuticals, where high-purity proteins are required for therapeutic use.
- ☆ It is also used in the development of diagnostic assays, where specific antibodies or other biomolecules are immobilized on a solid support to selectively capture and detect target molecules in a sample.
- ☆ In addition, it has applications in several other fields, including drug discovery, proteomics, genomics, and bioengineering. It has been used to identify and characterize the binding interactions between biomolecules, study protein-protein and protein-nucleic acid interactions, and develop novel materials and devices for biomedical applications.
- ☆ Affinity chromatography can be performed in various formats, including column chromatography, batch chromatography, and magnetic bead-based chromatography. The choice of format and ligand depends on the specific application and the properties of the target molecule.
- ☆ The method can also be optimized by using various strategies, such as gradient elution, pH optimization, or temperature optimization.

Note

- ☆ Affinity chromatography offers several advantages over other chroma-tography methods, such as high specificity, high selectivity, and high purity, making it an indispensable tool in various fields of biotechnology and biomedicine.
- ☆ It allows the isolation of proteins with high specificity, even in the presence of other closely related proteins.
- ☆ Sometimes, it is referred to as the absolute or the best method for purifying a protein.

2. Principles

2.1. Basic principles

- ☆ The principle of affinity chromatography is based on the fact that biomolecules have specific binding sites that interact with other molecules in a highly specific manner (Figure 1).

☆ The specific binding interaction between a target molecule and a ligand (or affinity matrix) immobilized on a solid support helps in separation. The ligand can be a biological macromolecule such as a protein, peptide, nucleic acid, carbohydrate, or small molecule that has a high affinity and selectivity for the target molecule.

☆ The binding affinity between a biomolecule and its ligand is influenced by several factors, including the chemical nature of the binding sites, the size and shape of the molecules, and the strength of the binding interaction. By immobilizing a ligand on a solid support, it is possible to selectively capture and isolate a biomolecule of interest from a complex mixture based on its specific binding affinity for the ligand (Figure 1).

☆ The basic principle of affinity chromatography involves four steps: immobilization of the ligand on a solid support, sample application, washing, and elution.

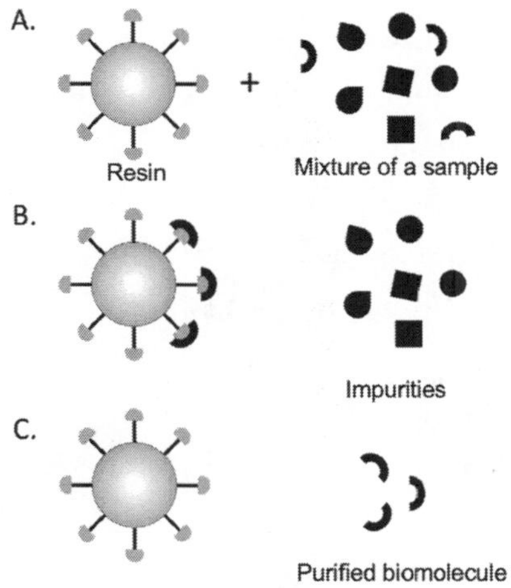

Figure 1: Basic principle of affinity chromatography. Affinity chromatography separates molecules based on specific interactions between a ligand on the resin and the target molecule in the sample.

Note

☆ One of the most common applications is the purification of recombinant proteins expressed in heterologous systems such as bacteria, yeast, or mammalian cells. In this case, the protein of interest is fused to a purification tag such as His-tag, GST-tag, FLAG-tag, or MBP-tag, which has a high affinity and specificity for a specific ligand immobilized on the column. After purification, the tag can be removed by proteolytic cleavage or by using a specific enzyme, leaving the target protein in its native form.

2.2. Types of interactions used in affinity chromatography

- ☆ Affinity chromatography relies on the specific binding interactions between a target molecule and a ligand immobilized on a solid support.
- ☆ The choice of interaction depends on the specific application and the properties of the target molecule.
- ☆ Understanding the different types of interactions used in affinity chromatography is critical for selecting the appropriate ligand and optimizing the chromatographic conditions for the desired purification outcome.
- ☆ Different types of interactions can be utilized in affinity chromatography, including reversible and irreversible binding interactions.
- ☆ Typical biological interactions used in affinity chromatography are listed in **Table 1**.

2.2.1. Reversible binding interactions

- ☆ Reversible binding interactions are the most common types of interactions used in affinity chromatography.
- ☆ These interactions involve the reversible association of the target molecule with the immobilized ligand.
- ☆ The most commonly used reversible interactions are based on hydrogen bonding, van der Waals forces, and electrostatic interactions.
- ☆ These interactions are typically weaker than irreversible interactions but can be optimized by adjusting the pH, temperature, and salt concentration of the buffer.

2.2.2. Irreversible binding interactions

- ☆ Irreversible binding interactions involve the covalent attachment of the target molecule to the immobilized ligand.
- ☆ These interactions are typically stronger and more stable than reversible interactions but can be less specific.
- ☆ Common irreversible interactions used in affinity chromatography include thiol-disulfide and amine-carboxyl interactions.

Table 1: Typical biological interactions used in affinity chromatography.

Types of ligand	Target molecule
Substrate analogue	Enzymes
Antibody	Antigen
Lectin	Polysaccharide
Nucleic acid	Complementary base sequence

Types of ligand	Target molecule
Hormone	Receptor
Avidin	Biotin/Biotin-conjugated molecule
Calmodulin	Calmodulin binding partner
Glutathione	GST fusion protein
Protein A or Protein G	Immunoglobulins
Nickel-NTA	Polyhistidine fusion protein

2.3. Factors affecting affinity chromatography

- ☆ The performance of affinity chromatography depends on several factors, including ligand selection, matrix selection, and elution conditions.
- ☆ Understanding these factors is essential for achieving successful and reproducible purifications. Careful consideration of ligand selection, matrix selection, and elution conditions, as well as other parameters, can optimize the performance of affinity chromatography for specific applications.

2.3.1. Ligand selection

- ☆ Ligand selection is critical for the success of affinity chromatography.
- ☆ The ligand must have a high affinity and selectivity for the target molecule while also being stable and compatible with the chromatographic conditions.
- ☆ The choice of ligand depends on the properties of the target molecule, such as its size, shape, charge, and hydrophobicity.

2.3.2. Matrix selection

- ☆ Matrix selection is another important factor in affinity chromatography performance.
- ☆ The matrix should have a high binding capacity, low non-specific binding, and mechanical stability.
- ☆ Commonly used matrices include agarose, cellulose, and silica-based materials.

2.3.3. Elution conditions

- ☆ Elution conditions also play a critical role in affinity chromatography performance.
- ☆ The elution buffer must be optimized to elute the target molecule while minimizing non-specific binding and denaturation. T
- ☆ The elution buffer can be adjusted by changing the pH, salt concentration, or using competitive inhibitors.

2.3.4. Other factors

- ☆ Other factors that can affect affinity chromatography performance include sample preparation, column packing, and flow rate.
- ☆ Sample preparation is critical to ensure that the target molecule is in the appropriate buffer conditions for chromatography.
- ☆ Proper column packing is necessary to ensure even flow and maximize binding capacity.
- ☆ Flow rate can affect the kinetics of the interaction between the target molecule and ligand and should be optimized for each specific application.

3. Resins in Affinity Chromatography

- ☆ The resin in affinity chromatography typically consists of three components: the matrix (or solid support), the linker, and the ligand (Figure 2).

1. **Matrix** (solid support) provides a stable and inert surface for immobilizing the linker and the ligand.
2. **Linker** connects the ligand to the solid support. It must be stable, hydrophilic, and able to withstand the purification conditions.
3. **Ligand** is a molecule that specifically binds to the target molecule and is immobilized onto the linker. The ligand is the specific capture agent for the target molecule during the purification process.

- ☆ The components of an affinity chromatography resin work together to provide a highly specific and efficient method for the purification of biomolecules.

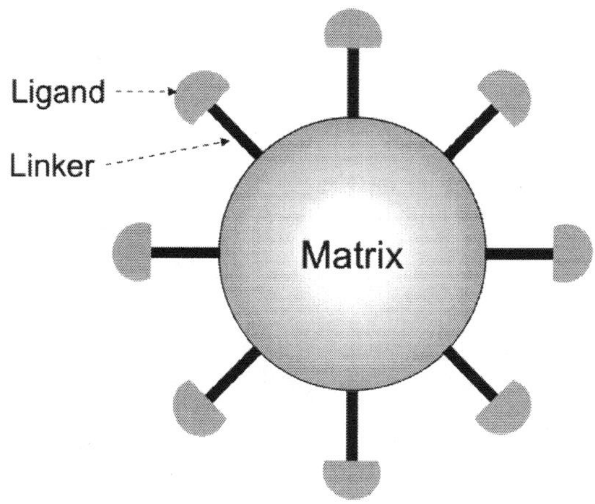

Figure 2: Components of affinity chromatography resin. The resin consists of a support matrix (solid support), a linker (spacer arm), and a ligand. These determine its specificity and binding capacity for target molecules.

3.1. Matrix (solid support)

- ☆ The matrix is a solid support (affinity support) onto which a ligand is immobilized through a linker.
- ☆ The matrix should be chemically and physically inert.
- ☆ The choice of the matrix is critical for the success of affinity chromatography, as it determines the efficiency and selectivity of the purification process.
- ☆ The matrix must be inert, stable, and able to immobilize the ligand without interfering with the binding of the target biomolecule.
- ☆ There are several types of matrices that can be used in affinity chromatography, including agarose, sepharose, cellulose, silica, and magnetic beads. Each matrix has its unique advantages and limitations.
- ☆ The choice of the matrix is based on the specific ligand and target biomolecule, as well as the intended application of the purified product.

3.1.1. Agarose

- ☆ Agarose is a widely used matrix in affinity chromatography due to its high mechanical strength, low cost, and easy availability.
- ☆ Agarose-based matrices are hydrophilic, and their pore size can be controlled by adjusting the degree of cross-linking.
- ☆ Agarose-based matrices are ideal for the immobilization of ligands that require gentle handlings, such as enzymes, antibodies, and lectins.

3.1.2. Sepharose

- ☆ Sepharose is another popular matrix for affinity chromatography, which is composed of cross-linked dextran and has a highly porous structure.
- ☆ Sepharose-based matrices are available in various pore sizes and are suitable for the immobilization of a wide range of ligands, including proteins, peptides, carbohydrates, and nucleic acids.
- ☆ They can be used for both batch and column chromatography, making them a versatile option for affinity purification.

3.1.3. Cellulose

- ☆ Cellulose-based matrices are another common choice for affinity chromatography, which are composed of microcrystalline cellulose fibers.
- ☆ Cellulose-based matrices are available in both bead and membrane forms and can be modified with various ligands, including carbohydrates, lectins, and enzymes.

- ☆ They are particularly useful for the purification of carbohydrates and glycoproteins.

3.1.4. Silica

- ☆ Silica-based matrices are a newer option in affinity chromatography, which are composed of porous silica particles coated with a hydrophilic polymer.
- ☆ Silica-based matrices are highly efficient and provide a high surface area for ligand immobilization.
- ☆ They are ideal for the purification of small molecules, such as peptides, oligonucleotides, and drugs.

3.1.5. Magnetic beads

- ☆ Magnetic beads are a unique type of matrix that has gained popularity in recent years, particularly for their ability to automate and streamline the purification process.
- ☆ Magnetic beads are composed of a magnetic core coated with a hydrophilic polymer. They are suitable for the immobilization of a wide range of ligands, including antibodies, enzymes, and nucleic acids.
- ☆ It can be performed in both batch and column formats, providing a high-throughput option for purification.

3.1.6. Other matrices

- ☆ In addition to these commonly used matrices, several specialized matrices are available for specific applications. For example, hydroxyapatite-based matrices are useful for the purification of phosphoproteins, while metal chelate-based matrices are useful for the purification of histidine-tagged proteins.

3.2. Linkers (spacer arm)

- ☆ The linker (spacer arm) is a critical component of an affinity chromatography resin, as it connects the ligand to the solid support.
- ☆ The linker must be stable, hydrophilic, and able to withstand the conditions used in the purification process. It also plays a role in determining the accessibility of the ligand to the target molecule.
- ☆ There are various types of linkers used in affinity chromatography, including hydrophilic polymers, peptides, and small organic molecules. The choice of linker depends on the nature of the ligand and the target molecule, as well as the intended use of the purified biomolecule.
- ☆ A few common linkers used in affinity chromatography are discussed below:

3.2.1. Polyethylene glycol (PEG)

- ☆ Polyethylene glycol (PEG) is a flexible and hydrophilic polymer that can be easily modified to introduce functional groups for conjugation with the ligand and the solid support.
- ☆ PEG linkers can provide good separation and high binding capacity and are often used to purify antibodies and enzymes.

3.2.2. Glutathione linker

- ☆ Another commonly used linker is the glutathione linker, which is used to purify glutathione-S-transferase (GST) fusion proteins.
- ☆ The linker consists of a tripeptide (Gly-Cys-Gly) that connects the GST tag to the solid support via a thiol group.
- ☆ This linker provides a stable and specific binding interaction between the GST tag and glutathione, enabling the efficient purification of GST fusion proteins.

3.2.3. Histidine linker

- ☆ The His-tag linker is among the most commonly used linker in affinity chromatography for the purification of proteins.
- ☆ The His-tag is a short peptide consisting of 6-10 histidine residues that is often fused to the N- or C-terminus of a recombinant protein.
- ☆ The His-tag linker consists of a short peptide sequence that connects the His-tag to the solid support.
- ☆ The linker typically comprises a flexible, hydrophilic polymer, such as PEG. It can be modified to introduce functional groups for conjugation with the solid support and the His-tag.
- ☆ The His-tag linker provides a stable and specific binding interaction between the His-tag and a metal ion, such as nickel or cobalt, immobilized on the solid support. This enables the efficient and selective purification of His-tagged proteins from a complex mixture.

Note

- ☆ The linker plays a crucial role in the design and performance of an affinity chromatography resin.
- ☆ The choice of linker depends on the specific requirements of the purification process and the characteristics of the ligand and the target molecule.

3.3. Ligands

- ☆ The ligand is the molecule that binds reversibly to a specific molecule or group of molecules, enabling purification by affinity chromatography.
- ☆ It provides the selectivity and specificity necessary for the efficient and effective purification of target molecules.
- ☆ The ligand interacts with the target molecule through specific binding sites or functional groups, allowing for the selective separation and isolation of the target molecule from a complex mixture.
- ☆ There are several types of ligands commonly used in affinity chromatography, including antibodies, enzymes, receptors, and small molecules. Ligands, characteristics, and elution conditions of a few affinity tags are shown in Table 2.
- ☆ Antibodies are frequently used as ligands due to their high specificity and affinity for their target antigen.
- ☆ Enzymes can also be used as ligands, as they can specifically recognize and bind to substrates or inhibitors.
- ☆ Receptors are used as ligands to isolate ligands or drugs that specifically bind to the receptor.
- ☆ Small molecules, such as metal ions, can also be used as ligands in affinity chromatography, allowing for the selective purification of target molecules based on metal-ion interactions. Metal ions are commonly used as ligands in immobilized metal ion affinity chromatography (IMAC) to purify histidine-tagged proteins.
- ☆ The choice of ligand for a particular purification process depends on the specific interaction between the ligand and the target molecule.
- ☆ Factors such as the size, charge, and hydrophobicity of the target molecule, as well as the availability and stability of the ligand, must be considered when selecting a ligand for affinity chromatography.

Table 2: Ligands, characteristics, and elution conditions of a few affinity tags.

Affinity tag (in target molecule)	Ligand	Characteristic	Elution condition
Histidine (His-tag)	Ni^{2+}-NTA, Co^{2+}-CMA (chelators)	Metal-affinity chromatography	Imidazole 20-250 mM or low pH
Glutathione S-transferase (GST-tag)	Glutathione	Increase solubility	5-10 mM reduced glutathione
Maltose-binding protein (MBP-tag)	Cross-linked amylose	Increase solubility	10 mM maltose

FLAG-tag	Anti-FLAG specific antibody	Non-denaturing purification, enterokinase cleavage site, and antibody-base purification	Competition with FLAG peptide, low pH elution with glycine, or 2-5 mM EDTA

Note

- ☆ The linker plays a crucial role in the design and performance of an affinity chromatography resin.
- ☆ The choice of linker depends on the specific requirements of the purification process and the characteristics of the ligand and the target molecule.
- ☆ The dissociation constant (k_d) for the ligand-target complex should ideally be in the range of 10^{-4} to 10^{-8} M in free solution.
- ☆ Interactions involving dissociation constants greater than 10^{-4} M are likely to be too weak for successful affinity chromatography. Conversely, if the dissociation constant is lower than 10^{-8} M, elution of the bound substance is likely to be difficult.

4. Methodology

- ☆ Affinity chromatography is a powerful method to isolate and purify specific biomolecules from complex mixtures based on their specific binding interactions with a ligand immobilized on a solid support.
- ☆ Before performing the affinity chromatography, the solid support is prepared. The solid support used in affinity chromatography is typically a porous matrix made of a material such as agarose, sepharose, or silica. The matrix is functionalized with a ligand that specifically binds to the target molecule. The ligand can be attached to the matrix by covalent or non-covalent interactions. The immobilization can be achieved by chemical cross-linking, adsorption, or affinity labelling.
- ☆ The procedure of affinity chromatography involves three steps, which are described below (Figure 3):

1. **Loading the sample:** The sample containing the target molecule is loaded onto the column containing the immobilized ligand. The sample is typically prepared in a buffer that is compatible with the binding interaction between the ligand and the target molecule. The target molecule interacts and binds to the ligand.
2. **Separation of the target molecule:** After the sample is loaded, the column is washed with a buffer to remove any unbound or non-specifically bound molecules. The target molecule is retained on the column due to its specific

binding to the ligand while the other molecules pass through. The strength of the binding interaction depends on the affinity and specificity of the ligand for the target molecule, as well as the chromatography conditions, such as pH, temperature, ionic strength, and solvent composition. The washing conditions depend on the properties of the ligand and the target molecule, as well as the strength of the binding interaction.

3. **Elution of the target molecule:** Elution is the final step in affinity chromatography, where the purified target molecule is eluted (recovered) from the column. To elute the target molecules, the buffer conditions are altered to reverse (weaken) the interaction between the target molecules and ligands. The choice of elution method depends on the affinity of the target molecule to the ligand and the strength of the interaction between them. There are several elution methods used in affinity chromatography, including competitive elution, pH elution, salt elution, and chaotropic elution.

   a) **Competitive elution** involves adding a competing ligand to the column to disrupt the interaction between the target molecule and the immobilized ligand (e.g., reduced glutathione, imidazole, etc.).

   b) **pH elution** involves changing the pH of the eluent to weaken the interaction between the target molecule and the immobilized ligand.

   c) **Salt elution** involves increasing the salt concentration of the eluent to weaken the interaction.

   d) **Chaotropic elution** involves using a chaotropic agent (denaturation agent) to disrupt the interaction between the target molecule and the immobilized ligand.

Note

☆ The elution buffer is collected in fractions, and the purity and identity of the target molecule are analysed by various methods such as gel electrophoresis, mass spectrometry, or enzyme assays.

☆ The elution method should be carefully considered to ensure the target molecule is effectively eluted while maintaining its integrity and purity.

☆ The elution conditions depend on the strength of the binding interaction between the ligand and the target molecule, as well as the stability of the target molecule under different conditions.

☆ After the target molecule is eluted, the resin is regenerated by washing with a buffer that removes any remaining contaminants or non-specifically bound molecules. The resin can be reused for further experiments by repeating the loading, separation, and elution steps.

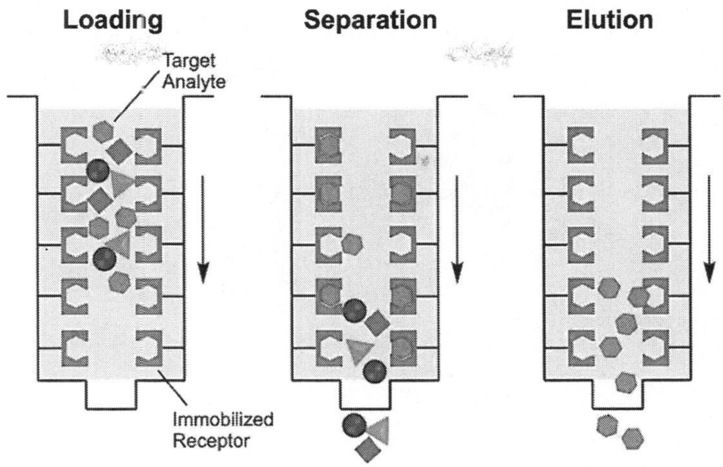

Figure 3: Steps in affinity chromatography. The procedure of affinity chromatography involves three basic steps: (i) loading the sample onto the column, (ii) separating the target molecule from the other components, and (iii) eluting the purified target molecule from the resin using a specific eluent.

Note

☆ The methodology of affinity chromatography can be further optimized by using various strategies such as gradient elution, pH optimization, or temperature optimization.

1. **Gradient elution** involves using a buffer with a gradually increasing concentration of a competing ligand or salt to selectively elute the target molecule.
2. **pH optimization** involves adjusting the pH of the buffer to optimize the binding interaction between the ligand and the target molecule.
3. **Temperature optimization** involves adjusting the temperature of the buffer to optimize the stability of the target molecule under different conditions.

Note

☆ Affinity chromatography can also be performed in different formats, such as batch, column, or magnetic bead-based formats.

1. **Batch affinity chromatography** involves mixing the immobilized ligand with the sample in a container, followed by separation of the bound and unbound molecules.

2. **Column affinity chromatography** involves packing the immobilized ligand into a column, followed by loading and elution of the sample.
3. **Magnetic bead-based affinity chromatography** involves immobilizing the ligand onto magnetic beads, followed by mixing with the sample and separation of the bound and unbound molecules using a magnetic field.

5. Types of Affinity Chromatography

5.1. Pseudo affinity chromatography

- ☆ Pseudoaffinity chromatography is a modification of traditional affinity chromatography that utilizes a pseudoaffinity ligand to capture and purify target molecules.
- ☆ Unlike traditional affinity chromatography, where the immobilized ligand has a specific binding site for the target molecule, pseudoaffinity chromatography relies on non-specific interactions between the target molecule and a modified ligand.
- ☆ Pseudoaffinity ligands are designed to interact with the target molecule through hydrophobic, electrostatic, or other non-specific interactions. This allows for the purification of a wider range of molecules than traditional affinity chromatography. The modified ligands can also be synthesized or purchased at a lower cost than traditional affinity ligands.
- ☆ Pseudoaffinity chromatography has been successfully used to purify various biomolecules, including proteins, nucleic acids, and carbohydrates. It has also been used to purify synthetic compounds and natural products.
- ☆ One of the key advantages of pseudoaffinity chromatography is its versatility. The non-specific nature of the interactions between the target molecule and the modified ligand allows for the purification of a wider range of molecules with varying properties. Pseudoaffinity chromatography can also be easily scaled up for industrial applications.

5.2. Dye-ligand chromatography

- ☆ Dye-ligand chromatography is a type of affinity chromatography that uses dyes as the immobilized ligands to selectively capture and purify target molecules.
- ☆ Dye-ligands are typically derived from synthetic dyes that have been modified to have specific chemical groups that can interact with the target molecule.

- ☆ The dye-ligand can selectively bind to the target molecule through a combination of electrostatic, hydrophobic, and/or hydrogen bonding interactions. The target molecule can then be eluted by changing the pH or ionic strength of the buffer.
- ☆ Dye-ligand chromatography has been successfully used to purify various biomolecules, including proteins, nucleic acids, and carbohydrates. It has also been used to purify synthetic compounds and natural products.
- ☆ One of the key advantages of dye-ligand chromatography is its high selectivity and specificity. The chemical modifications of the dye-ligands allow for the selective capture and purification of target molecules. Dye-ligand chromatography also has a high binding capacity, which makes it useful for large-scale purification.
- ☆ However, there are some limitations to dye-ligand chromatography. The non-specific nature of the interactions between the target molecule and the dye-ligand can sometimes lead to the co-purification of unwanted molecules. Additionally, the cost of synthesizing or purchasing dye-ligands can be high.

6. Applications

- ☆ Affinity chromatography has many applications in various fields of biochemistry and biotechnology. Some of the major applications of affinity chromatography are:

6.1. Purification of proteins

- ☆ Affinity chromatography is widely used for the purification of proteins from complex mixtures. A specific ligand is immobilized on a solid support and used to selectively capture the target protein from the mixture.
- ☆ The most common ligands used for protein purification are antibodies, enzymes, and lectins.
- ☆ The purified protein can be further characterized by various methods such as gel electrophoresis, mass spectrometry, or enzyme assays.

6.2. Purification of synthetic compounds

- ☆ Affinity chromatography is not limited to the purification of biomolecules but can also be used to purify synthetic compounds.
- ☆ Synthetic molecules can be designed with specific functional groups that can interact with the immobilized ligands on the chromatographic matrix. This allows for the selective purification of synthetic compounds with high purity and yield.

- Affinity chromatography can be used to purify small molecules, such as drugs and pharmaceutical intermediates, as well as larger synthetic molecules, such as polymers and nanoparticles.
- The use of affinity chromatography in the purification of synthetic compounds has become an increasingly popular and efficient method in the chemical industry.

6.3. Enzyme immobilization

- Affinity chromatography can be used to immobilize enzymes on a solid support. The enzyme is covalently or non-covalently attached to the ligand immobilized on the support.
- The immobilized enzyme can be used for various applications such as biocatalysis, biosensors, or drug discovery.

6.4. Drug discovery

- Affinity chromatography can be used to identify drug candidates by screening small molecule libraries against a specific protein target.
- The protein is immobilized on a solid support, and the library of small molecules is applied to the column. The compounds that bind specifically to the target protein are retained on the column, while the non-specific molecules are washed away.
- The bound compounds can be eluted and further characterized for their binding affinity and specificity.

6.5. Diagnostics

- Affinity chromatography can be used to isolate and purify specific biomolecules such as antibodies, antigens, or nucleic acids for diagnostic applications.
- The purified biomolecule can be used as a reagent in various diagnostic tests such as ELISA, Western blot, or PCR.

6.6. Protein-protein interactions

- Affinity chromatography can be used to study protein-protein interactions by capturing one protein on a column and using it as bait to capture its interacting partner from a complex mixture.
- The interacting protein can be further characterized by various techniques such as mass spectrometry or protein sequencing.

6.7. Glycoprotein analysis

- Affinity chromatography can be used to isolate and analyse glycoproteins, which are proteins with covalently attached carbohydrate chains.
- Lectins, which are proteins that bind specifically to certain carbohydrate structures, can be used as ligands to isolate and purify glycoproteins.

6.8. Environmental and industrial applications

- In environmental applications, affinity chromatography can be used to selectively remove pollutants, such as heavy metals, from contaminated water sources.
- In industrial applications, affinity chromatography can be used for protein purification in the food and beverage industry, as well as for the production of high-value chemicals and pharmaceuticals.

08 Hydrophobic Interaction Chromatography

CHAPTER SUMMARY

- Hydrophobic interaction chromatography (HIC) is a separation method based on the reversible interaction between hydrophobic ligands on the stationary phase and hydrophobic regions of target molecules, enabling their purification and analysis.
- HIC relies on the differential hydrophobicity of molecules to separate them. It utilizes a hydrophobic stationary phase that interacts with hydrophobic regions of target molecules, allowing for their retention and elution based on their hydrophobic properties.
- In HIC, the sample is loaded onto a hydrophobic stationary phase, and separation is achieved by exploiting the differences in hydrophobic interactions between the sample components and the stationary phase. Elution is typically achieved by decreasing the hydrophobicity of the mobile phase.
- HIC is commonly used in protein purification, separation of biomolecules, and removal of impurities.
- HIC and reverse phase chromatography (RPC) are both chromatographic methods based on hydrophobic interactions, but differ in the nature of the stationary phase. HIC utilizes hydrophobic ligands, while RPC employs a nonpolar stationary phase bonded to silica or polymer.

1. Introduction

- Hydrophobic interaction chromatography (HIC) is a powerful method used for separating proteins, peptides, and other biomolecules based on their hydrophobic properties.
- This method relies on the interaction between the hydrophobic groups of the target molecule and the hydrophobic ligands of the stationary phase in the column.
- In HIC, the stationary phase consists of hydrophobic ligands, such as alkyl chains or aryl groups, which interact with the hydrophobic patches on the target molecule's surface. The mobile phase typically consists of a high salt

concentration buffer, which competes with the hydrophobic interactions and elutes the target molecule.

- ☆ HIC was first introduced in the early 1970s as a separation method for proteins and peptides. It was developed by Robert K. Scopes, who noticed that the addition of high salt concentrations to protein solutions resulted in protein precipitation. This led to the development of HIC, which utilized the same principle to separate proteins based on their hydrophobicity. Since then, HIC has become a popular method in the field of biochemistry and is widely used for protein purification.
- ☆ HIC is a valuable tool in the purification and analysis of biomolecules, particularly proteins and peptides. HIC is often used in conjunction with other chromatographic methods, such as ion exchange chromatography and size exclusion chromatography, to achieve high-purity samples.
- ☆ Some common applications of HIC include the purification of enzymes, monoclonal antibodies, and recombinant proteins. HIC is also used in the analysis of protein-ligand interactions and protein folding studies.
- ☆ HIC is a gentle separation method that does not denature or damage the target molecule, making it suitable for the purification of delicate proteins and peptides. HIC is highly selective, as it separates molecules based on their hydrophobic properties, which can be specific to each target molecule. It is also a cost-effective method, as the stationary phase can be regenerated and reused multiple times.
- ☆ However, HIC is a low-resolution method, as it separates molecules based on their hydrophobicity, which can be influenced by various factors, such as pH and temperature. It is also not suitable for the separation of molecules that have similar hydrophobic properties, as they may not be resolved from each other. In addition, HIC requires the use of high salt concentrations, which can affect the stability of the target molecule and may require additional steps to remove the salt from the purified sample.

2. Principles

- ☆ HIC exploits the hydrophobic interactions between a hydrophobic stationary phase and a target biomolecule. The principle behind HIC is based on the fact that hydrophobic groups tend to aggregate in aqueous solutions, which is the basis of protein precipitation. By using hydrophobic ligands in the stationary phase, HIC separates proteins, peptides, and other biomolecules based on their varying degrees of hydrophobicity.

2.1. Stationary phase

- ☆ The stationary phase in HIC is usually made up of hydrophobic ligands attached to a matrix, such as agarose or silica beads.

- ☆ The most common hydrophobic ligands used in HIC are alkyl chains, such as octyl and butyl, or aryl groups, such as phenyl and benzyl.
- ☆ The length of the alkyl chain or the type of aryl group used determines the strength of the hydrophobic interactions with the target molecule.
- ☆ The stationary phase must be compatible with the sample being analysed and should not cause any significant denaturation or aggregation of the target molecule.

2.2. Mobile phase

- ☆ The mobile phase in HIC typically contains a high salt concentration, which reduces the strength of the hydrophobic interactions between the stationary phase and the target molecule.
- ☆ The salt concentration must be optimized for each target molecule to ensure efficient separation.
- ☆ A buffer solution is usually used to control the pH of the mobile phase, which can also affect the strength of the hydrophobic interactions.

2.3. Sample preparation

- ☆ Prior to HIC, the sample must be prepared to remove any contaminants and reduce the complexity of the mixture.
- ☆ Common methods used for sample preparation include size exclusion chromatography and ion exchange chromatography.
- ☆ The sample is then applied to the column, which has been equilibrated with the mobile phase.
- ☆ The sample should be in a buffer that is compatible with the mobile phase to prevent any precipitation or aggregation.

2.4. Separation process

- ☆ The separation process in HIC is based on the differential hydrophobicity of the target molecule compared to other components in the sample.
- ☆ The sample is applied to the column and interacts with the hydrophobic ligands in the stationary phase. The more hydrophobic components of the sample will interact more strongly with the stationary phase, resulting in delayed elution from the column. Less hydrophobic components will elute more quickly.

2.5. Elution

- ☆ The elution of the target molecule from the column is accomplished by changing the conditions of the mobile phase.

☆ Elution is done by decreasing the salt concentration, using an organic solvent, or changing the pH.

☆ The elution profile of the target molecule can be monitored using a UV-visible spectrophotometer or other analytical techniques.

☆ The target molecule can be collected in fractions and analysed for purity and activity.

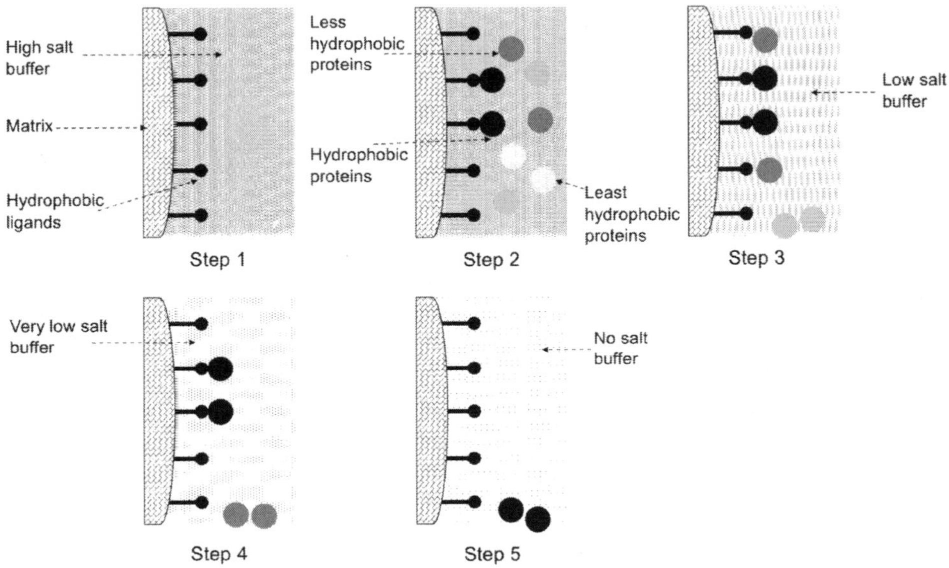

Figure 1. HIC procedure: Biomolecules are applied to a hydrophobic stationary phase. Water mediates hydrophobic interactions between the target molecules and the stationary phase, resulting in selective binding. Elution is achieved by decreasing hydrophobicity through the addition of high salt or organic solvent concentrations. The target molecules are eluted in order of increasing hydrophobicity, enabling their separation and purification.

Note

☆ **Role of water in HIC:** The role of water in HIC is crucial for the successful separation of target molecules.

☆ Water acts as a mediator in HIC by enabling the formation of the hydration layer, which allows for selective binding and separation of hydrophobic biomolecules based on their relative hydrophobicity.

☆ In HIC, a hydrophobic stationary phase is used. The hydrophobic interactions occur between the non-polar regions of the biomolecules and the stationary phase. Water plays a critical role in this process by mediating the interactions between the hydrophobic stationary phase and the hydrophobic regions of the target molecules.

- ☆ Water molecules form a structured layer around hydrophobic surfaces due to the strong cohesive forces between water molecules. This layer, known as the hydration layer, shields the hydrophobic groups from direct contact with the hydrophobic stationary phase. As a result, the target molecules are retained on the column based on their relative hydrophobicity, with more hydrophobic molecules binding more strongly to the stationary phase.
- ☆ The addition of a high concentration of salts or organic solvents to the mobile phase disrupts the hydration layer, weakening the hydrophobic interactions and promoting the elution of the target molecules. Thus, water is essential for maintaining the hydrophobic interactions, as it facilitates the reversible binding of the biomolecules to the stationary phase.

3. Types of HIC

- ☆ There are various variants of HIC.
- ☆ Column chromatography is the most commonly used type of HIC for protein purification, while high-throughput screening (HTS) HIC is useful for high-throughput screening. Hydrophobic charge induction chromatography (HCIC) is a variant of HIC that utilizes charged ligands in addition to hydrophobic ligands, making it a powerful tool for separating proteins with similar hydrophobic properties but different charges.

3.1. Column chromatography

- ☆ Column chromatography is the most commonly used type of HIC.
- ☆ This method involves packing a column with a hydrophobic stationary phase, such as octyl or phenyl ligands attached to a matrix, and passing the sample through the column. The more hydrophobic components of the sample will interact more strongly with the stationary phase, resulting in delayed elution from the column. Less hydrophobic components will elute more quickly.
- ☆ The elution of the target molecule is accomplished by changing the conditions of the mobile phase, such as decreasing the salt concentration or changing the pH.
- ☆ Column chromatography can be used for small-scale or large-scale protein purification and is a widely used method in the field of biochemistry and biotechnology.

3.2. High-throughput screening (HTS) HIC

- ☆ High-throughput screening (HTS) HIC is a method that uses microtiter plates to screen many samples simultaneously.
- ☆ It utilizes hydrophobic beads in the wells of a microtiter plate to bind target molecules. The sample is added to the wells, and the hydrophobic interactions between the target molecule and the beads are exploited to achieve separation. The plate is washed to remove non-bound components, and the target molecule is eluted using a low salt buffer.
- ☆ HTS HIC is a valuable tool for screening large libraries of compounds or proteins for potential drug candidates or other applications.

3.3. Hydrophobic charge induction chromatography (HCIC)

- ☆ Hydrophobic charge induction chromatography (HCIC) is a variant of HIC that utilizes charged ligands in the stationary phase in addition to hydrophobic ligands.
- ☆ The charged ligands induce a charge on the target molecule, which enhances the hydrophobic interactions with the stationary phase.
- ☆ HCIC is often used for the separation of proteins that have similar hydrophobic properties but different charges, such as protein variants or isoforms.
- ☆ The sample is applied to the column and interacts with the charged and hydrophobic ligands in the stationary phase. The more hydrophobic and charged components of the sample will interact more strongly with the stationary phase, resulting in delayed elution from the column. The elution of the target molecule is accomplished by changing the conditions of the mobile phase, such as decreasing the salt concentration or changing the pH.
- ☆ HCIC is a powerful tool for protein separation and purification, particularly for complex mixtures with similar hydrophobic properties.

4. Methodology

- ☆ To perform HIC, a series of steps are required, including selecting a resin for the column, packing the column, preparing the mobile phase, preparing the sample, and elution.
- ☆ The selection of the appropriate parameters for each step is critical for achieving efficient separation of the target protein.

4.1. Resin selection

- ☆ The selection of the appropriate resin is a critical step in performing HIC.
- ☆ The choice of resin depends on the hydrophobicity of the target protein and the desired separation conditions.

☆ Typical hydrophobic resins include phenyl or octyl groups attached to a matrix, such as agarose or silica beads.

☆ Resins with higher hydrophobicity will interact more strongly with the target protein, leading to greater separation.

4.2. Column packing

☆ The column packing step involves packing the resin into a column with a specific diameter and length.

☆ The column packing density and uniformity are critical parameters that affect the separation efficiency of HIC.

☆ The resin is packed into the column using a slurry of the resin in a low salt buffer. The column is then washed to remove any air pockets, and the resin is packed to the desired density.

4.3. Mobile phase preparation

4.3.1 Salt concentration

☆ The salt concentration in the mobile phase is a critical parameter in HIC.

☆ High salt concentrations will disrupt the hydrophobic interactions between the protein and the stationary phase, leading to lower separation efficiency.

☆ The salt concentration has to be adjusted to optimize the separation of the target protein.

4.3.2. pH adjustment

☆ The pH of the mobile phase can also affect the hydrophobic interactions between the protein and the stationary phase.

☆ The pH can be adjusted to optimize the separation of the target protein.

4.4. Sample preparation

4.4.1. Protein concentration

☆ The protein concentration in the sample is an important parameter in HIC.

☆ The protein concentration should be high enough to ensure that enough target protein interacts with the stationary phase for efficient separation.

4.4.2. Buffer Exchange

☆ The buffer in which the protein is dissolved can also affect the separation efficiency of HIC.

☆ The buffer must be exchanged to a low salt buffer to reduce the ionic strength of the sample and improve the hydrophobic interactions between the protein and the stationary phase.

4.5 Elution

4.5.1. Step elution

- ☆ In step elution, the mobile phase is changed to a lower salt concentration or a different pH to elute the target protein.
- ☆ This is done in a series of steps to elute the target protein with increasing efficiency.

4.5.2. Gradient elution

- ☆ In gradient elution, the salt concentration or pH of the mobile phase is changed gradually to elute the target protein.
- ☆ This method can result in improved separation efficiency compared to step elution.

5. Applications

- ☆ HIC is a versatile method that has a wide range of applications in the field of biochemistry and biotechnology.
- ☆ The hydrophobic interactions between the target biomolecules and the hydrophobic stationary phase make HIC an effective method for separating and purifying a variety of biomolecules.

5.1. Protein purification

- ☆ **Antibody purification:** HIC is widely used for the purification of monoclonal and polyclonal antibodies. Antibodies are typically purified from complex matrices such as cell culture supernatants or serum. HIC can effectively separate the target antibody from other proteins present in the sample due to the differences in their hydrophobicity.
- ☆ **Enzyme purification:** HIC is a common method used for the purification of enzymes. Enzymes are often purified from complex mixtures such as cell lysates or fermentation broth. HIC can effectively separate enzymes based on their hydrophobicity, allowing for the efficient purification of the target enzyme.

5.2. Separation of other biomolecules

- ☆ **DNA and RNA separation:** HIC is also used for the separation of DNA and RNA based on their hydrophobicity. HIC can effectively separate single-stranded and double-stranded DNA and RNA fragments.
- ☆ **Separation of lipids and polysaccharides:** HIC is also used for the separation of lipids and polysaccharides. HIC is an effective method for the separation of lipids due to their hydrophobic nature. Polysaccharides can also be separated based on their hydrophobicity, making HIC an effective method for their purification.

- ☆ **Purification of viruses and viral vectors:** HIC is commonly used for the purification of viruses for vaccine production. HIC can effectively separate viruses based on their hydrophobicity, allowing for the efficient purification of the target virus. HIC is also used for the purification of viral vectors for gene therapy. HIC can effectively separate viral vectors from other impurities in the sample.
- ☆ **DNA-protein complex purification:** HIC is also used for the purification of DNA-protein complexes. HIC can effectively separate DNA-bound proteins from unbound proteins in the sample, allowing for the efficient purification of the target complex.

5.3. Industrial applications

- ☆ **Biopharmaceutical industry:** HIC is widely used in the biopharmaceutical industry for the purification of therapeutic proteins. HIC is often used in conjunction with other chromatography methods, such as ion exchange chromatography and size exclusion chromatography, to achieve highly purified proteins.
- ☆ **Food and beverage industry:** HIC is also used in the food and beverage industry for the purification of food additives and flavours.

6. Reverse Phase Chromatography

- ☆ HIC and reverse phase chromatography (RPC) are two related but distinct methods used for separating biomolecules. The main difference between the two is the type of stationary phase used. In HIC, the stationary phase is a hydrophobic material, while in RPC, the stationary phase is a hydrophobic material that is chemically bonded to a polar support.
- ☆ RPC is a powerful separation method widely used in the purification of peptides and proteins. In RPC, a hydrophobic stationary phase, such as C18, is used in conjunction with a polar mobile phase, typically composed of water and organic solvents, such as acetonitrile or methanol. Proteins bind to the stationary phase through hydrophobic interactions, with the degree of binding influenced by the hydrophobicity of the protein and the properties of the stationary phase.
- ☆ Elution of the bound proteins is achieved by increasing the concentration of the organic solvent in the mobile phase, thereby reducing the hydrophobic interactions between the proteins and the stationary phase. The resulting elution profile can be analysed by various techniques, such as UV absorbance, mass spectrometry, or fluorescence, to identify the proteins and monitor the purification process.

☆ RPC has several advantages over other chromatographic methods, including high resolution, high selectivity, and high capacity. RPC is particularly useful for the separation of closely related proteins, such as different isoforms, and can be used for preparative scale protein purification. Furthermore, RPC is compatible with a wide range of protein samples, including those containing detergents or other additives.

7. Comparison of HIC and RPC

☆ HIC and RPC are two widely used chromatographic methods for the separation and purification of biomolecules.

☆ Both methods utilize hydrophobic interactions between the stationary phase and the sample molecules to achieve separation, but there are several key differences between the two (Table 1).

☆ RPC is generally more selective and efficient than HIC, but the choice between the two methods depends on the protein's characteristics, desired level of purity, and downstream applications.

☆ Understanding the similarities and differences between HIC and RPC is important for selecting the optimal purification strategy for a given protein.

Table 1: Table showing the differences between HIC and RPC.

HIC	RPC
Used for separating large, complex proteins or protein mixtures, as it can separate proteins based on subtle differences in hydrophobicity.	Used for high-throughput separations of small molecules.
Separates proteins based on reversible interactions between non-polar regions of proteins and hydrophobic groups on the stationary phase.	Uses a hydrophobic ligand on the stationary phase and elutes proteins by changing the polarity of the solvent system.
Utilizes a stationary phase composed of a hydrophilic material (e.g., agarose) that is derivatized with hydrophobic groups (e.g., butyl or phenyl groups).	Utilizes a hydrophobic stationary phase (e.g., C18 or C8) that interacts with the hydrophobic regions of the sample molecules.
Sample is loaded onto the column in a high salt buffer, which promotes the formation of hydrophobic interactions between the stationary phase and the hydrophobic regions of the sample molecules. The strength of these interactions can be modulated by changing the salt concentration, allowing for selective elution of the sample molecules.	Sample is loaded onto the column in a high organic solvent buffer (e.g., acetonitrile or methanol) and is eluted using a gradient of decreasing organic solvent concentration. The strength of the hydrophobic interactions between the stationary phase and the sample molecules is affected by the polarity of the mobile phase, making RPC particularly useful for the separation of non-polar and moderately polar molecules.

09 Gas Chromatography

CHAPTER SUMMARY

- Gas chromatography (GC) is used to separate and analyse volatile compounds. It relies on the differential interaction of analytes with a stationary phase and a mobile gas phase for separation.
- It operates on the principle of partitioning, where the sample is vaporized and separated based on its distribution between a stationary phase (a liquid or solid) and a carrier gas. The different compounds in the sample travel through the column at different rates, leading to their separation and detection.
- There are different types of GC methods, each offering unique advantages and is used for specific applications in analytical chemistry and research.
- GC finds extensive applications in various fields such as pharmaceuticals, environmental analysis, forensics, food and beverage industry, petrochemicals, and quality control. It is used for analysing volatile compounds, separating complex mixtures, and quantifying analytes with high sensitivity and selectivity.

1. Introduction

- Gas chromatography (GC) is an analytical method used for the separation and analysis of volatile compounds in a gaseous or vaporized state.
- It relies on the principles of partitioning and adsorption to achieve the separation of individual components in a complex mixture.
- GC is widely employed in various fields such as pharmaceuticals, environmental analysis, forensics, petrochemicals, and food and beverage industries.
- The development of GC is attributed to Archer J.P. Martin and Richard L.M. Synge, who were awarded the Nobel Prize in Chemistry in 1952 for their contributions. They introduced the concept of partition chromatography, which formed the foundation for modern GC.

- ☆ The evolution of GC technology continued with advancements in column technology, detectors, and computerized data analysis systems. The introduction of capillary columns in the 1960s revolutionized the field by offering higher separation efficiency and increased sample throughput. Further advancements led to the development of different detector types, including flame ionization detectors (FID), thermal conductivity detectors (TCD), electron capture detectors (ECD), and mass spectrometers (MS) for enhanced sensitivity and selectivity.
- ☆ The advantages of GC are :
  - ☆ **High separation efficiency:** GC provides excellent separation of complex mixtures due to its long and narrow capillary columns, allowing for the identification and quantification of individual components.
  - ☆ **Sensitivity and selectivity:** GC offers high sensitivity and selectivity in compound detection. Different detectors, such as FID, TCD, ECD, and MS, provide varying levels of sensitivity and can target specific analytes based on their properties.
  - ☆ **Wide range of applications:** GC is applicable to a broad range of compounds, making it suitable for various industries such as pharmaceuticals, environmental analysis, petrochemicals, and food and beverage.
- ☆ The disadvantages of GC are :
  - ☆ **Limited to volatile compounds:** GC is suitable only for volatile compounds that can be vaporized without decomposition. Non-volatile or thermally labile compounds may not be amenable to GC analysis.
  - ☆ **Sample preparation:** GC often requires complex and time-consuming sample preparation procedures, such as extraction and derivatization, to convert analytes into a volatile form suitable for analysis.
  - ☆ **Lack of structural information:** GC provides limited structural information about analytes. Additional techniques like mass spectrometry may be needed for compound identification.

2. Principles

- ☆ GC is widely used for separating and analysing volatile compounds in various fields such as chemistry, pharmaceuticals, environmental science, and forensic analysis. It enables the separation and analysis of complex mixtures with high resolution and sensitivity.

- It relies on the principles of partitioning and differential migration of analytes between a stationary phase and a flowing mobile phase.
- The efficiency of GC relies on the optimization of various parameters, including column dimensions, temperature programming, flow rates, and detection parameters.
- Understanding the principles and factors influencing GC is crucial for obtaining reliable and accurate analytical results.

2.1. Stationary phase

- In GC, the stationary phase is typically a high-boiling liquid or a solid adsorbent coated onto the inner walls of a capillary column.
- The choice of stationary phase depends on the analytes of interest and their interactions with the stationary phase.
- Common stationary phases include non-polar phases such as polydimethylsiloxane (PDMS) for non-polar analytes and polar phases such as polyethylene glycol (PEG) for polar analytes.

2.2. Mobile phase

- The mobile phase in GC is an inert carrier gas, most commonly helium or nitrogen.
- The role of the mobile phase is to carry the sample components through the column.
- The choice of carrier gas depends on factors such as cost, availability, and compatibility with the detection system.
- The mobile phase should be free from impurities that may interfere with the analysis.

2.3. Sample injection

- The sample injection process involves introducing a small volume of the sample into the chromatographic system.
- Common injection techniques include split injection, where a fraction of the sample is directed to the column, and splitless injection, where the entire sample is injected.
- The injection technique should be optimized to ensure accurate and reproducible results.

2.4. Separation

- The separation occurs as the sample components interact with the stationary phase while being carried by the mobile phase.

- ☆ The analytes partition between the stationary phase and the mobile phase based on their affinity for each phase.
- ☆ The degree of separation is determined by factors such as the boiling points, polarity, and molecular weight of the analytes, as well as the selectivity of the stationary phase.

2.5. Detection

- ☆ Detection in GC involves measuring the concentration of separated analytes as they elute from the column.
- ☆ Various detectors can be used in GC, including flame ionization detectors (FID), thermal conductivity detectors (TCD), electron capture detectors (ECD), and mass spectrometry (MS). Each detector offers different sensitivity, selectivity, and detection limits, allowing for the identification and quantification of analytes.

3. Types of GC

- ☆ GC is a versatile method that encompasses various types, each offering unique advantages and applications.

3.1. Gas-liquid chromatography

- ☆ Gas-liquid chromatography (GLC), also known as gas-liquid partition chromatography, is a widely used method for the separation and analysis of volatile and semi-volatile compounds.
- ☆ It relies on the partitioning of analytes between a gaseous mobile phase and a liquid stationary phase.
- ☆ GLC offers high resolution, sensitivity, and versatility, making it an indispensable tool in various scientific disciplines.
- ☆ The principle of GLC is based on the differential affinity of analytes for the liquid stationary phase and the gaseous mobile phase. The stationary phase is typically a liquid coating applied to the inner walls of a fused silica capillary column or a solid support material packed into a column. The choice of stationary phase depends on the specific analytes and their interactions with the stationary phase. Non-polar stationary phases, such as polydimethylsiloxane (PDMS), are used for separating non-polar compounds, while polar stationary phases, like polyethylene glycol (PEG), are used for polar compounds.
- ☆ The mobile phase in GLC is an inert gas, most commonly helium or nitrogen, which carries the analyte molecules through the column. The mobile phase is selected based on factors such as cost, availability, and compatibility with the detector. The carrier gas should be chemically inert and free from impurities that may interfere with the analysis.

☆ The separation process in GLC occurs as the sample components partition between the stationary phase and the mobile phase. Analytes that have a stronger affinity for the stationary phase will spend more time interacting with it, resulting in a slower elution time, while analytes with a weaker affinity will elute faster. The separation is influenced by factors such as the boiling points, polarity, molecular weight, and chemical structure of the analytes, as well as the selectivity of the stationary phase.

☆ Detection in GLC is typically performed using a variety of detectors, including FID, TCD, ECD, and MS. Each detector offers different sensitivity, selectivity, and detection limits, allowing for the identification and quantification of analytes.

☆ The GLC can be performed in either packed column GLC or capillary column GLC (Figure 1).

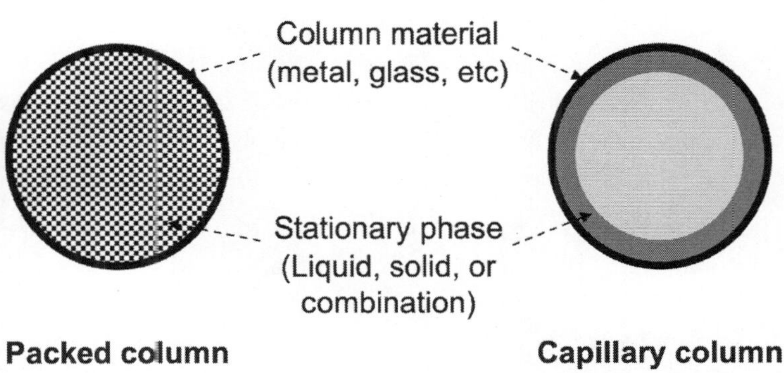

Figure 1: Comparison of packed column GLC and capillary column GLC. Packed column GLC utilizes a packed stationary phase for compound separation, while capillary column GLC employs a thin film stationary phase coated on a capillary tube.

3.1.1 Packed column GLC

☆ Packed column GLC utilizes a packed column filled with a solid support material coated with a liquid stationary phase.

☆ The stationary phase interacts with the analytes, leading to their separation based on their affinity for the stationary phase.

☆ Packed columns offer higher sample capacity but lower efficiency compared to capillary columns.

☆ It can be used for both GLC and GSC.

3.1.2. Capillary column GLC

☆ Capillary column GLC employs a fused silica capillary column coated with a stationary phase.

- ☆ Capillary columns provide higher resolution and efficiency due to their smaller diameter and longer length. They can withstand higher temperatures and allow for faster analysis.
- ☆ Capillary column GLC is suitable for analysing a wide range of compounds, including volatile and semi-volatile organic compounds.
- ☆ It can be used only for GLC.

3.2. Gas-solid chromatography

- ☆ Gas-solid chromatography (GSC) is a type of adsorption chromatography used for the separation and analysis of volatile compounds. It relies on the interaction of analytes with a solid stationary phase.
- ☆ GSC is particularly useful for the analysis of non-polar and moderately polar compounds.
- ☆ In GSC, the stationary phase is a solid material, typically packed into a column. The solid phase is selected based on its ability to adsorb analytes of interest. Common solid phases include activated charcoal, silica gel, and molecular sieves. The choice of stationary phase depends on the analytes being analysed and their interactions with the solid support.
- ☆ The separation process in GSC occurs as the analytes interact with the solid stationary phase. The analytes are adsorbed onto the solid surface based on their affinity for the stationary phase. The degree of adsorption depends on various factors, including the analyte's polarity, molecular weight, and chemical structure. Compounds with a stronger affinity for the solid phase will be retained longer, while those with weaker interactions will elute faster.
- ☆ The mobile phase in GSC is an inert carrier gas, such as helium or nitrogen, which flows through the column, carrying the analytes. The carrier gas transports the analytes through the column and does not participate in the separation process. The choice of carrier gas depends on factors such as availability, cost, and compatibility with the detection system.
- ☆ Detection in GSC can be accomplished using TCD, FID, and MS. These detectors measure the concentration of the analytes as they elute from the column, allowing for their identification and quantification.
- ☆ GSC is particularly useful for the separation of non-polar and moderately polar compounds that are not easily separated by other chromatographic methods. It is often used to analyse volatile organic compounds, hydrocarbons, and other non-polar species.

3.3. Gas-supercritical fluid chromatography

☆ Gas-supercritical fluid chromatography (GSFC) is a chromatographic method that utilizes supercritical fluids as the mobile phase for the separation and analysis of compounds.

☆ Supercritical fluids are substances that are maintained above their critical temperature and pressure, where they exhibit unique properties combining the characteristics of both liquids and gases. The most commonly used supercritical fluid in GSFC is carbon dioxide (CO_2).

☆ In GSFC, the supercritical fluid is used as a mobile phase that interacts with the analytes and transports them through the column. The separation is achieved based on the differential solubility of the analytes in the supercritical fluid. By adjusting the temperature and pressure, the solubility of the analytes can be modified, enabling their selective separation.

☆ GSFC offers several advantages over other chromatographic methods. First, supercritical fluids have low viscosities, which results in efficient mass transfer and faster analysis times. Second, the solubility of analytes in supercritical fluids can be easily adjusted by changing the temperature and pressure, allowing for fine-tuning of the separation. Third, supercritical fluids are highly compatible with a wide range of compounds, including non-polar, polar, and thermally labile analytes. Additionally, GSFC is an environmentally friendly method as it uses carbon dioxide, which is non-toxic and readily available.

☆ GSFC finds applications in various fields, such as pharmaceutical analysis, natural product analysis, environmental analysis, and food and beverage testing. It is particularly useful for the separation of compounds that are difficult to analyse by other chromatographic methods, such as thermally labile and non-volatile compounds. GSFC can also be coupled with other detection techniques like MS or UV-Vis spectroscopy for compound identification and quantification.

3.4. Capillary gas chromatography

☆ Capillary gas chromatography (CGC) utilizes a capillary column for the separation and analysis of volatile and semi-volatile compounds.

☆ It offers improved resolution, sensitivity, and peak capacity compared to other GC methods.

☆ The principle of CGC is based on the separation of analytes in a capillary column packed with a stationary phase. The stationary phase is coated onto the inner walls of a fused silica capillary, providing a large surface area for efficient separation. The choice of stationary phase depends on the specific analytes and their interactions with the coating material. Common stationary phases include polydimethylsiloxane (PDMS), polyethylene glycol (PEG), and cyanopropylphenyl.

- ☆ The separation process in CGC occurs as the analytes interact with the stationary phase. The separation is based on differences in boiling points, polarity, molecular weight, and chemical structure. Analytes with stronger interactions with the stationary phase will elute more slowly, while those with weaker interactions will elute faster, resulting in the separation of the compound mixture.
- ☆ Detection in CGC is typically performed using various detectors, such as FID, TCD, and MS. These detectors measure the concentration of the analytes as they elute from the column, enabling their identification and quantification.
- ☆ CGC offers several advantages over other chromatographic methods. The use of a capillary column allows for increased resolution and peak capacity due to the small diameter and high efficiency of the column. It also requires less sample volume, resulting in improved sensitivity. Additionally, CGC provides better separation for complex mixtures and can handle a wide range of analyte types.
- ☆ It has applications in various scientific fields and is essential for the analysis of complex mixtures. CGC has become a crucial tool in pharmaceuticals, environmental analysis, forensics, and food and flavour analysis.

4. Instrumentation

- ☆ The success of GC heavily relies on the instrumentation used, which comprises various components and systems designed to ensure accurate and reliable results.
- ☆ A schematic diagram illustrating the main components of GC instrumentation is shown in Figure 2.

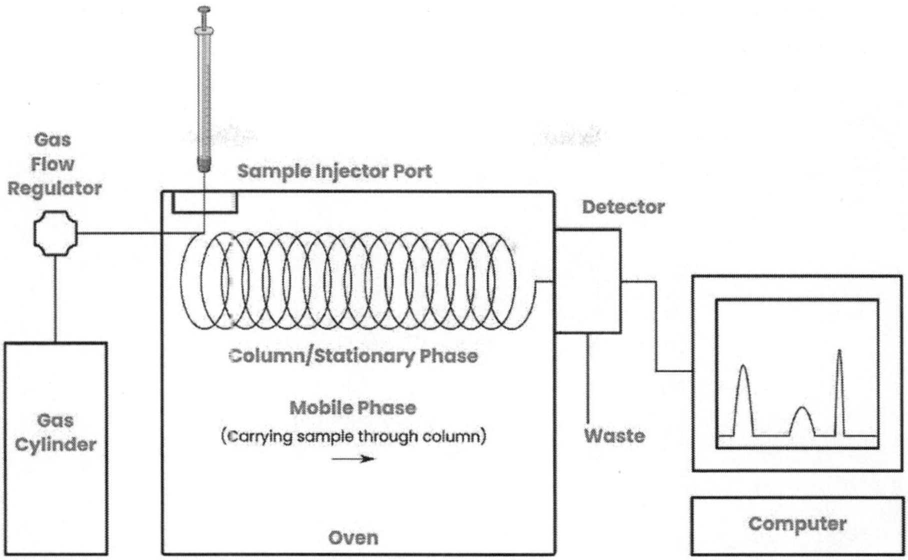

Figure 2: Schematic diagram illustrating the main components of GC instrumentation. The system includes a column for compound separation, a gas supply system for carrier gas delivery, a sample introduction system for injecting samples into the column, and a detector for compound detection. Together, these components enable accurate analysis of compounds in gas chromatography.

4.1. GC columns

- ☆ **Packed columns:** Packed columns are one type of GC column. They are filled with a solid support material coated with a liquid stationary phase. Packed columns are typically used for the analysis of complex mixtures and larger molecules.
- ☆ **Capillary columns:** Capillary columns are another type of GC column. They consist of a fused silica tube coated with a stationary phase or contain a stationary phase bonded to the inner wall. Capillary columns provide higher resolution and efficiency compared to packed columns and are suitable for the analysis of volatile and semi-volatile compounds.

4.2. Gas supply and delivery system

1. **Carrier gas:** The carrier gas is an essential component of the gas supply system. It is responsible for carrying the sample components through the column. Commonly used carrier gases include helium, nitrogen, and hydrogen. The choice of carrier gas depends on factors such as compatibility with the detector and the analytes being analysed.

2. **Make-up gas:** The make-up gas is used in certain detection systems to enhance the detector's performance and stability. It is introduced into the detector system to optimize the signal response and maintain a stable baseline.

4.3. Sample introduction system

1. **Split injection:** Split injection is a common sample introduction technique in GC. It involves splitting the injected sample into two portions, with only a small portion entering the column for analysis. The split ratio can be adjusted to control the amount of sample reaching the column.
2. **Splitless injection:** Splitless injection is another sample introduction technique that allows the entire sample to enter the column. This technique is suitable for trace analysis or when analytes are present at low concentrations.

4.4. Detector system

1. **Flame ionization detector (FID):** The flame ionization detector is widely used in GC. It operates by burning the analyte in a hydrogen-air flame, producing ions that can be measured as an electrical current. FID is highly sensitive and offers a wide dynamic range.
2. **Thermal conductivity detector (TCD):** The thermal conductivity detector measures changes in the thermal conductivity of the carrier gas caused by the presence of analytes. It is a non-destructive and universal detector that can be used for a wide range of compounds.
3. **Electron capture detector (ECD):** The electron capture detector is a sensitive and selective detector used in GC. It measures the reduction in electric current caused by analytes with electron-capturing properties. ECDs are commonly used for the detection of compounds such as halogenated compounds, pesticides, and PCBs.
4. **Mass spectrometer detectors:** Mass spectrometer detectors are crucial components in gas chromatography-mass spectrometry (GC-MS) systems. These detectors play a vital role in analysing the ions produced by the mass spectrometer, providing valuable information about the compounds being analysed. Common mass spectrometer detectors include electron ionization (EI), chemical ionization (CI), and selected ion monitoring (SIM). They enable sensitive and selective identification and quantification of analytes in complex mixtures. The choice of the detector depends on the specific analytical requirements and the nature of the compounds being studied.

4.5. Data analysis and processing

- ☆ **Chromatogram interpretation:** Chromatogram interpretation involves analysing the peaks and patterns produced by the separated components. It includes determining retention times, peak identification, and assessing peak shapes for qualitative analysis.
- ☆ **Quantitative analysis:** Quantitative analysis involves determining the concentration of the analytes in a sample. It requires calibration curves, peak integration, and data processing algorithms to accurately quantify the analytes.

5. Applications

- ☆ GC finds extensive applications in various fields, including environmental analysis, food and beverage industry, pharmaceuticals, petrochemicals, environmental forensics, natural product analysis, and forensic science.
- ☆ Its versatility, high sensitivity, and ability to separate complex mixtures make it an indispensable analytical tool for qualitative and quantitative analysis of volatile and semi-volatile compounds, ensuring the safety, quality, and compliance of diverse products and environments.

5.1. Environmental analysis

- ☆ GC is extensively employed in environmental analysis for the detection and quantification of pollutants in air, water, soil, and sediment samples.
- ☆ It can identify and measure volatile organic compounds (VOCs), polycyclic aromatic hydrocarbons (PAHs), pesticides, herbicides, and other contaminants.
- ☆ GC is crucial in environmental monitoring, assessing pollution levels, and ensuring compliance with regulatory standards.

5.2. Food and beverage analysis

- ☆ GC is commonly used in the food and beverage industry for quality control, authenticity testing, and flavour analysis.
- ☆ It can detect and quantify volatile compounds responsible for aroma and flavour, such as terpenes in essential oils, aldehydes in coffee, and esters in fruits.
- ☆ GC is also employed for the analysis of food contaminants, including pesticides, mycotoxins, and additives, ensuring the safety and integrity of food products.

5.3. Pharmaceutical analysis

- ☆ GC plays a significant role in pharmaceutical analysis, particularly in drug development, quality control, and pharmacokinetic studies.
- ☆ It is used to analyse drug compounds, identify impurities, determine drug stability, and assess bioavailability.
- ☆ GC is also utilized in forensic toxicology to detect drugs of abuse and analyse biological samples such as blood, urine, and hair.

5.4. Petrochemical analysis

- ☆ GC is widely employed in the petrochemical industry for the analysis of hydrocarbon mixtures and for determining the composition of petroleum products.
- ☆ It is utilized for the characterization of fuels, crude oil, lubricants, and other petroleum derivatives.
- ☆ GC is valuable in assessing the quality, purity, and compliance of these products with industry specifications.

5.5. Environmental forensics

- ☆ GC is an essential tool in environmental forensics, which involves investigating and identifying the source of pollutants and contaminants in environmental samples.
- ☆ It helps determine the presence of specific markers or unique compounds that can link the source of pollution to a particular location or activity, aiding in environmental remediation and legal investigations.

5.6. Forensic science

- ☆ GC is widely utilized in forensic science for the analysis of trace evidence, including arson investigation, explosive residue analysis, and drug identification.
- ☆ It helps in identifying and quantifying volatile compounds associated with criminal activities, providing crucial evidence in legal proceedings.

5.7. Natural product analysis

- ☆ GC is used to analyse natural products and essential oils for their chemical composition, purity, and authenticity.
- ☆ It enables the identification and quantification of bioactive compounds, such as terpenoids, alkaloids, and phenols, present in plant extracts and herbal formulations.
- ☆ GC is vital in the quality control of herbal medicines, perfumes, and cosmetic products.

10 High-Performance Liquid Chromatography

CHAPTER SUMMARY

- High-performance liquid chromatography (HPLC) is a powerful method used to separate, identify, and quantify components of complex mixtures using pressure.
- It operates on the principles of differential partitioning of analytes between a stationary phase and a mobile phase, allowing for efficient separation and analysis of compounds.
- HPLC can be performed as reversed-phase, normal-phase, ion-exchange, size exclusion, and affinity chromatography, each employing different stationary and mobile phase combinations for specific separation requirements.
- HPLC column selection and optimization involve considering factors such as stationary phase chemistry, particle size, column length, and mobile phase composition to achieve optimal separation and efficiency in chromatographic analysis.
- HPLC instrumentation and operation involve components like pumps, injectors, columns, detectors, and data acquisition systems, along with parameters such as flow rate, mobile phase composition, and temperature control, for efficient separation and analysis of analytes.
- HPLC finds applications in various fields such as pharmaceuticals, environmental analysis, food and beverage industry, forensics, and research, enabling precise separation and analysis of complex mixtures.
- HPLC and fast-pressure liquid chromatography (FPLC) are two related but distinct chromatographic methods with different principles, applications, and target analytes. While HPLC is used for small molecule analysis, FPLC is specifically designed for protein purification and characterization.

1. Introduction

- High-performance liquid chromatography (HPLC), also called high-pressure liquid chromatography, is a powerful analytical method used for the separation, identification, and quantification of chemical compounds in complex mixtures.

- It is based on the principles of chromatography, where a sample is dissolved in a mobile phase and passed through a stationary phase.
- HPLC utilizes a liquid mobile phase and a column packed with a stationary phase, which allows for the separation of compounds based on their physicochemical properties.
- The development of HPLC can be traced back to the early 20th century when chemists began exploring ways to improve the efficiency and resolution of liquid chromatography methods. In the 1960s, the introduction of small-particle, porous packings and high-pressure pumps revolutionized liquid chromatography, leading to the birth of HPLC.
- Initially, HPLC was primarily used in the pharmaceutical industry for drug analysis and quality control. However, its applications quickly expanded to various fields, including environmental analysis, food and beverage analysis, forensic sciences, and biochemistry. The continuous advancements in technology, column materials, and detector systems have further enhanced the capabilities of HPLC, making it an indispensable tool in modern analytical laboratories.
- HPLC offers several advantages over other chromatographic methods. Firstly, it provides excellent separation power, allowing for the separation of complex mixtures with high resolution and sensitivity. The wide range of available stationary phases and mobile phase compositions provides versatility in separation methods, making it suitable for a diverse range of applications.
- HPLC can analyse a wide range of compounds, including polar and non-polar analytes, small molecules, and large biomolecules. It is also capable of quantitative analysis, enabling accurate determination of compound concentrations.
- Another advantage of HPLC is its wide range of detection options, including UV-Vis absorbance, fluorescence, electrochemical detection, and mass spectrometry. This enables the identification and quantification of compounds with high selectivity and sensitivity.
- However, HPLC does have some limitations. One major disadvantage is its relatively slow analysis time compared to other chromatographic methods. The need for equilibration and longer separation times can be a drawback, particularly in high-throughput analysis. HPLC also requires specialized equipment and skilled personnel, making it relatively expensive to establish and maintain. The cost of consumables, such as columns and solvents, can also be significant.

2. Principles

- ☆ HPLC is based on fundamental principles that involve a stationary phase, mobile phase, sample introduction, separation process, and detection. A simplistic procedure of HPLC is shown in Figure 1.
- ☆ The stationary phase, typically packed in a column, interacts with the sample molecules to facilitate separation.
- ☆ The mobile phase, a liquid solvent or solvent mixture, carries the sample through the column.
- ☆ Sample introduction ensures accurate and reproducible results.
- ☆ The separation occurs as analytes interact with the stationary phase based on differential affinity.
- ☆ Detection of separated analytes is achieved using various detectors, such as UV-Vis, fluorescence, or mass spectrometry.
- ☆ HPLC enable the separation, identification, and quantification of compounds in complex mixtures with high precision and accuracy, making it an indispensable method in analytical chemistry.

2.1. Stationary phase

- ☆ The stationary phase in HPLC is a critical component that interacts with the sample molecules to facilitate their separation. It is typically a solid or a porous material packed into a column.
- ☆ The choice of stationary phase depends on the nature of the compounds being analyzed and the separation mechanism desired.
- ☆ Commonly used stationary phases include silica-based materials, such as C18 (octadecylsilane), C8 (octylsilane), and C4 (butylsilane), which provide hydrophobic interactions. Other stationary phases, such as ion-exchange resins or chiral selectors, are utilized for specific applications requiring different separation mechanisms.

2.2. Mobile phase

- ☆ The mobile phase in HPLC is a liquid solvent or a combination of solvents that carries the sample through the column. The selection of an appropriate mobile phase is crucial for achieving efficient separation. It should possess suitable properties, such as appropriate polarity and compatibility with the stationary phase.
- ☆ Commonly used mobile phases include mixtures of water and organic solvents, such as acetonitrile or methanol. The ratio of water to organic solvent can be adjusted to optimize the separation, taking into account

the hydrophobicity and solubility of the analytes. Buffer solutions may also be added to control the pH or provide ion-pairing effects for certain separations.

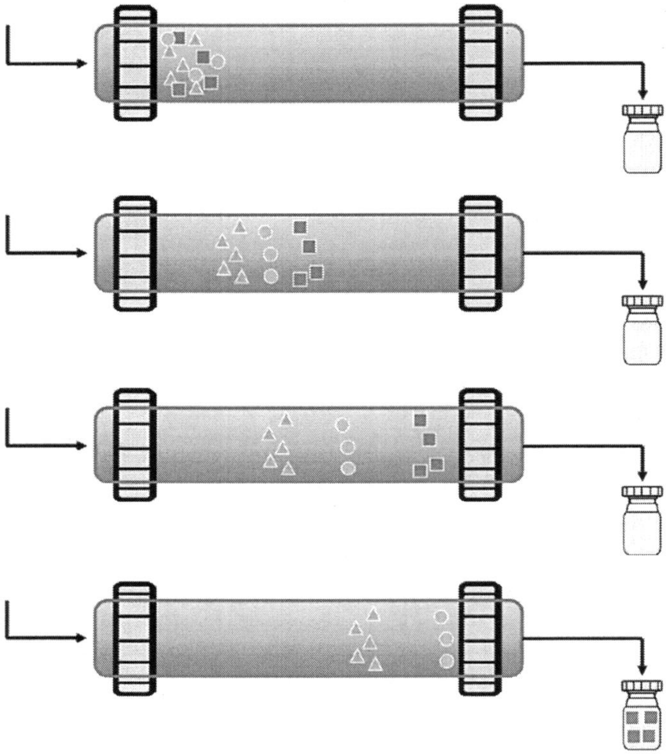

Figure 1: Movement of molecules in an HPLC column. The HPLC procedure involves sample injection, mobile phase flow through the column, separation of analytes based on their interactions with the stationary phase, detection of separated analytes, and generation of chromatograms for analysis and quantification.

2.3. Sample injection

- ☆ Sample injection is the process of introducing the analyte mixture into the HPLC system. It is essential to ensure accurate and reproducible results.
- ☆ The most common method of sample introduction in HPLC is through an autoinjector or manual injection using a syringe.
- ☆ Prior to injection, the sample may need to undergo pre-treatment steps, such as filtration or dilution, to remove particulate matter and adjust the analyte concentration within the linear range of detection. Depending on the nature of the analyte, additional sample preparation methods, such as extraction or derivatization, may be required.

2.4. Separation

- ☆ The separation process in HPLC occurs as the sample components interact with the stationary phase. The separation is based on the differential affinity between the analyte molecules and the stationary phase, resulting in different retention times and elution profiles.
- ☆ Various separation mechanisms are employed in HPLC, including adsorption, partition, ion exchange, size exclusion, and affinity. The choice of separation mechanism depends on the nature of the analytes and the desired selectivity.
- ☆ During the separation process, the mobile phase continuously flows through the column, carrying the sample components. The analytes that interact more strongly with the stationary phase will have longer retention times, while those with weaker interactions will elute faster. This differential retention leads to the separation of the sample components.

2.5. Detection

- ☆ Detection is the final stage of the HPLC process, where the separated analytes are detected and quantified. A wide range of detectors can be used in HPLC, depending on the nature of the analytes and the desired sensitivity.
- ☆ UV-Vis absorbance detectors are commonly employed and provide broad applicability, as many compounds absorb light in the ultraviolet or visible region. Fluorescence detectors offer higher sensitivity and selectivity for analytes with fluorescent properties. Other detection techniques, such as refractive index detection, electrochemical detection, and mass spectrometry, are also utilized for specific applications.
- ☆ The choice of the detector depends on factors such as the compound's properties, concentration range, and required sensitivity.
- ☆ Multiple detectors can be used in series or in combination to enhance detection capabilities and obtain complementary information.

3. Types of HPLC

- ☆ HPLC encompasses various types that are utilized for different separation purposes.
- ☆ Different chromatographic methods can be performed using the HPLC platform to increase their high resolution and sensitivity.
- ☆ HPLC can be run in the 'normal phase' or 'reverse phase', depending on the polarity of the compounds.

- ☆ Different types of HPLC provide valuable tools for separating and analyzing a wide range of compounds.

3.1. Reverse-phase chromatography HPLC

- ☆ Reverse-phase chromatography (RPC) is one of the most commonly used modes of HPLC. In RPC, the stationary phase is non-polar, while the mobile phase is polar. This configuration allows for the separation of compounds based on their hydrophobicity.
- ☆ The stationary phase in RPC is typically composed of hydrophobic materials, such as C18 (octadecylsilane), C8 (octylsilane), or C4 (butylsilane). These materials contain long hydrocarbon chains that interact with hydrophobic regions of analytes. The more hydrophobic compounds will have stronger interactions with the stationary phase, resulting in longer retention times.
- ☆ RPC is suitable for a wide range of analytes, including non-polar and moderately polar compounds. It is commonly used for pharmaceutical analysis, drug discovery, environmental monitoring, and lipid analysis.

3.2. Normal-phase chromatography HPLC

- ☆ Normal-phase chromatography (NPC), also known as adsorption chromatography, is the opposite of RPC in terms of the polarity of the stationary and mobile phases. In NPC, the stationary phase is polar, while the mobile phase is non-polar.
- ☆ The stationary phase in NPC is typically a polar material, such as silica gel or alumina. The polar stationary phase interacts with polar functional groups of analytes through adsorption, leading to their separation. Compounds with higher polarity will have stronger interactions and longer retention times.
- ☆ NPC is particularly useful for separating highly polar and hydrophilic compounds, such as small organic acids, amino acids, and carbohydrates. It is also employed for the separation of stereoisomers and chiral compounds.

3.3. Ion-exchange chromatography HPLC

- ☆ Ion-exchange chromatography (IEC) utilizes ion-exchange resins as the stationary phase to separate analytes based on their charge properties. In IEC, the stationary phase contains charged functional groups that interact with ions present in the analyte mixture.
- ☆ The stationary phase can be either cation-exchange or anion-exchange, depending on the desired separation. Cation-exchange resins have negatively charged functional groups and retain positively charged ions or molecules, while anion-exchange resins have positively charged functional groups and retain negatively charged species.

- ☆ IEC is widely used in the analysis of biological samples, such as proteins, peptides, nucleic acids, and pharmaceutical formulations. It allows for the separation and purification of charged compounds based on their differences in charge and ionic strength.

3.4. Size exclusion chromatography HPLC

- ☆ Size exclusion chromatography (SEC) is a type of HPLC that separates analytes based on their size or molecular weight. The stationary phase in SEC consists of porous beads with defined pore sizes.
- ☆ As the analyte mixture is introduced into the column, smaller molecules or analytes that can access the internal pores of the beads will be temporarily trapped, resulting in longer retention times. In contrast, larger molecules or analytes that cannot enter the pores will pass through the column more quickly.
- ☆ SEC is commonly used for the separation of polymers, proteins, peptides, and nanoparticles. It is an effective method for determining the molecular weight distribution of macromolecules and estimating their sizes.

3.5. Affinity chromatography HPLC

- ☆ Affinity chromatography is a specialized type of HPLC that exploits the specific interactions between a target analyte, and a selective ligand immobilized on the stationary phase. The ligand is designed to selectively bind to the analyte of interest while allowing other components to pass through the column.
- ☆ Affinity chromatography is highly specific and is widely used for the purification, isolation, and analysis of proteins, enzymes, antibodies, and other biomolecules. The ligands used in affinity chromatography include antibodies, enzymes, lectins, and other molecules that can recognize and bind to specific target analytes.
- ☆ The binding between the ligand and analyte is reversible, allowing for the elution of the target molecule under appropriate conditions. Affinity chromatography enables the isolation of proteins with high purity and specificity, making it a valuable tool in biochemical research, biotechnology, and pharmaceutical development.

4. HPLC column selection and optimization

- ☆ The selection of an appropriate HPLC column is crucial for achieving efficient and accurate separations.
- ☆ Factors such as column dimensions, materials, packing, temperature, and pressure must be carefully considered to ensure optimal performance.

- ☆ Proper selection and optimization of HPLC columns enable the analysis of a wide range of compounds and support high-quality results in various scientific and analytical applications.

4.1. Column dimensions and materials

- ☆ Column dimensions, such as length, inner diameter, and particle size, play a significant role in the efficiency and resolution of the chromatographic process.
- ☆ The column length affects the separation efficiency, with longer columns providing better resolution but at the cost of longer analysis times.
- ☆ The inner diameter influences the flow rate and the amount of sample required, with smaller diameters resulting in higher backpressure and slower elution.
- ☆ The particle size of the stationary phase affects the efficiency of the separation, with smaller particles offering higher resolution but also higher backpressure.
- ☆ In addition to dimensions, the choice of column material is important. Common column materials include silica, polymer-based materials, and hybrid materials. Silica-based columns are versatile and suitable for a wide range of applications. Polymer-based columns offer enhanced stability and compatibility with certain sample matrices. Hybrid materials combine the advantages of both silica and polymer-based materials.

4.2. Column packing and efficiency

- ☆ Proper column packing is essential for achieving optimal separation efficiency in HPLC. The packing material, typically silica particles, should be evenly distributed and tightly packed to minimize channelling and ensure efficient analyte interactions.
- ☆ Factors influencing column packing include particle size, particle shape, and packing pressure. Smaller particle sizes provide higher efficiency but also increase backpressure. Uniformly shaped particles lead to more efficient packing, while irregular shapes can cause voids and inefficiencies. Packing pressure ensures proper settling of the particles and helps achieve a homogeneous packing bed.
- ☆ Efficiency can also be influenced by column chemistry and bonding phases. Different stationary phases, such as C18 (octadecylsilane), C8 (octylsilane), or ion-exchange materials, offer specific separation characteristics. Bonding phases can be modified to enhance selectivity, improve peak shape, or increase compatibility with specific sample matrices.

4.3. Column temperature and pressure

- ☆ Column temperature and pressure are important parameters in HPLC optimization.
- ☆ Temperature affects the analyte interactions with the stationary phase and the mobile phase viscosity, thus influencing separation efficiency and selectivity. Higher temperatures can reduce analysis times and improve peak shapes, but they must be within the thermal stability limits of the column and analytes.
- ☆ Pressure, or backpressure, is a critical consideration in HPLC column optimization. It affects the flow rate, resolution, and column lifetime. High backpressure can result in poor efficiency, peak broadening, and increased wear on the system. Therefore, selecting an appropriate particle size, column dimensions, and mobile phase flow rate is necessary to maintain optimal pressure conditions.
- ☆ Optimization of column temperature and pressure is typically achieved through trial and error experiments, considering the specific requirements of the separation and the limitations of the HPLC system.

5. Instrumentation

- ☆ HPLC relies on sophisticated instrumentation to facilitate the separation and analysis of compounds.
- ☆ The key components of HPLC instrumentation include pumps, injectors, detectors, and data acquisition systems. A schematic diagram of basic HPLC is shown in Figure 2.

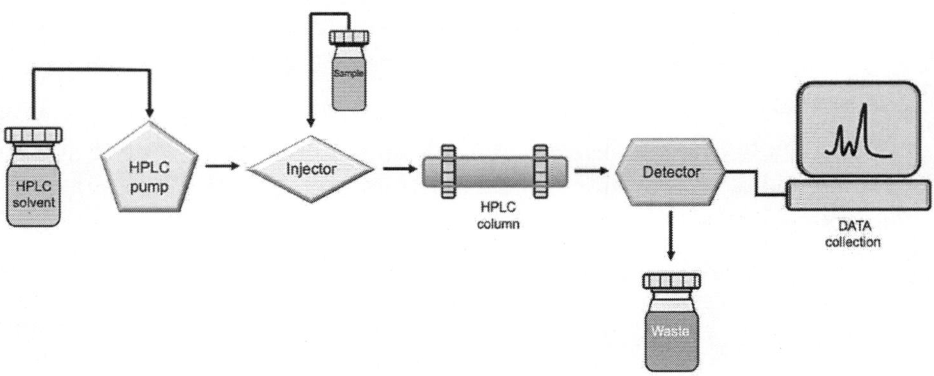

Figure 2: HPLC instrumentation. The HPLC system consists of pumps, injectors, detectors, and data acquisition systems. Pumps generate the high-pressure gradient, injectors introduce the sample, detectors detect analytes, and data acquisition systems collect and process signals for analysis.

5.1. Pumps and injectors

- ☆ The HPLC system consists of several key components that work together to facilitate the separation and analysis of compounds.
- ☆ Pumps are responsible for generating the high-pressure gradient necessary to push the mobile phase through the column. Modern HPLC systems often utilize high-pressure, binary or quaternary gradient pumps, which allow precise control of the mobile phase composition and flow rate.
- ☆ Injectors are used to introduce the sample into the mobile phase stream. They are equipped with a sample loop where a specific volume of the sample is loaded. The sample is then injected into the mobile phase stream, initiating the separation process.

5.2. Detectors and data acquisition

- ☆ Detectors play a crucial role in HPLC analysis, as they detect and quantify the separated analytes.
- ☆ UV-Visible detectors are commonly used in HPLC and rely on the absorption of UV or visible light by the analytes. They provide high sensitivity and allow for the analysis of a wide range of compounds.
- ☆ Fluorescence detectors utilize the fluorescence emission of certain compounds when excited by specific wavelengths of light. They offer higher sensitivity than UV-Visible detectors, making them suitable for trace analysis and the detection of compounds with low UV absorbance.
- ☆ Mass spectrometers (MS) coupled with HPLC, known as LC-MS, provide high sensitivity, selectivity, and structural information about the analytes. They are particularly useful for the identification and quantification of complex mixtures and compounds with low UV absorbance.
- ☆ Data acquisition systems collect and process the detector signals, generating chromatograms and peak data. Modern HPLC systems often feature advanced software for real-time monitoring, data analysis, and result interpretation.

5.3. Auto-samplers and sample preparation

- ☆ Auto-samplers are devices that automate the injection of samples into the HPLC system. They improve accuracy, precision, and throughput by precisely controlling the injection volume and minimizing manual handling. Auto-samplers are commonly used in routine analysis and high-throughput applications.
- ☆ Sample preparation is an essential step in HPLC analysis. It involves extracting, purifying, and concentrating the analytes from complex

matrices. Techniques such as solid-phase extraction (SPE) and liquid-liquid extraction (LLE) are often employed to prepare samples for HPLC analysis, ensuring accurate and reliable results.

5.4. HPLC method development and validation

- ☆ HPLC method development is a critical process that involves optimizing various parameters to achieve efficient separation of analyses.
- ☆ It includes the selection of appropriate columns, mobile phase composition, and chromatographic conditions.
- ☆ Method development aims to achieve selectivity and specificity, ensuring accurate identification and quantification of target analyses.
- ☆ Once a method is developed, it must be validated to ensure its reliability and suitability for its intended purpose. Method validation involves a comprehensive evaluation of various performance parameters such as linearity, accuracy, precision, sensitivity, and robustness.
- ☆ Linearity assesses the relationship between analyze concentration and detector response.
- ☆ Accuracy measures the closeness of results to the true value.
- ☆ Precision examines the consistency and reproducibility of the method.
- ☆ Sensitivity determines the minimum detectable concentration of analytes.
- ☆ Robustness evaluates the method's ability to withstand small variations in parameters without a significant impact on results.
- ☆ Method validation is essential for ensuring the reliability, accuracy, and reproducibility of analytical results. It helps to establish the performance characteristics of the method and ensures that it meets the requirements of regulatory guidelines and quality standards.
- ☆ Proper method development and validation are crucial. A well-developed and validated HPLC method provides a solid foundation for reliable and accurate analysis.

6. Fast-pressure liquid chromatography

- ☆ Fast-pressure liquid chromatography (FPLC) is a powerful method used for the separation, purification, and analysis of proteins.
- ☆ It is a variant of liquid chromatography specifically designed for protein-based applications.
- ☆ FPLC utilizes a range of separation mechanisms, including ion exchange, affinity, and size exclusion chromatography, to achieve high-resolution protein separations.

- ☆ FPLC systems are equipped with specialized columns and chromatography media tailored for protein interactions.
- ☆ With FPLC, proteins can be purified from complex mixtures, characterized based on their properties, and further studied for various applications in biochemistry, biotechnology, and pharmaceutical research.
- ☆ FPLC offers advantages such as flexibility, scalability, and ease of use, making it an indispensable tool in protein analysis and purification workflows.

7. Difference between HPLC and FPLC

- ☆ HPLC and FPLC are both widely used methods for the separation and analysis of compounds, but they differ in several aspects.
- ☆ HPLC is primarily utilized for the separation of small organic molecules based on their chemical properties and interactions with a stationary phase. On the other hand, FPLC is specifically designed for protein separation and purification.
- ☆ FPLC utilizes specialized chromatography media and columns tailored for protein interactions, allowing for efficient separation based on size, charge, or affinity.
- ☆ The differences between HPLC and FPLC lie in their separation principles, stationary phases, sample preparation requirements, flow rates, pressures, and applications. Understanding these distinctions is essential for selecting the appropriate method for specific analytical needs. The key differences between HPLC and FPLC are listed in Table 1.

Table 1: Key differences between HPLC and FPLC.

	HPLC	FPLC
Principle of separation	Primarily used for the separation of small organic molecules. It relies on the differential partitioning of analytes between a mobile phase and a stationary phase. The separation is based on differences in polarity, size, and chemical properties of the analytes.	Specifically designed for the separation of proteins and biomolecules. It utilizes various separation mechanisms, including ion exchange, affinity, and size exclusion, to achieve efficient separation based on the unique properties of proteins.
Stationary phase and column	Employs a wide range of stationary phases, such as reversed phase, normal-phase, ion exchange, and size exclusion columns. The choice of the column depends on the properties of the analytes and the separation goal.	Utilizes specialized chromatography media specifically designed for protein separation. The stationary phase commonly consists of resins or beads functionalized with ligands specific to protein interactions, such as charged groups for ion exchange or affinity ligands for specific binding.

	HPLC	FPLC
Sample preparation	Sample preparation is required to extract and purify analytes before analysis. This may involve extraction, filtration, and derivatization steps to convert the analytes into a suitable form for separation.	Typically used for the analysis of purified proteins or complex mixtures with minimal sample preparation. Protein samples are often purified using methods like centrifugation, filtration, or chromatography prior to FPLC analysis.
Flow rate and pressure	Operate at higher flow rates and require higher pressure to achieve efficient separations. This is due to the smaller particle sizes and longer columns used in HPLC.	Operate at lower flow rates and lower pressures compared to HPLC. This is because protein separations are more sensitive to shear forces, and higher pressure can lead to denaturation or damage to the proteins.
Applications	Applications in various industries, including pharmaceuticals, environmental analysis, food and beverage, and forensic sciences. It is suitable for the separation and analysis of small organic molecules, drugs, and compounds with diverse chemical properties.	Specifically designed for protein purification and analysis. It finds applications in biochemistry, biotechnology, proteomics, and pharmaceutical research, where the separation and characterization of proteins and biomolecules are of primary interest.

8. Applications

- ☆ HPLC is a versatile analytical method with numerous applications in various fields.
- ☆ Its ability to separate and analyse complex mixtures makes it an invaluable tool in pharmaceuticals, environmental analysis, food and beverage, forensic sciences, biotechnology, petrochemicals, clinical diagnostics, research, and academia.
- ☆ HPLC enables accurate, sensitive, and reliable analysis, contributing to advancements in scientific knowledge, quality control, and regulatory compliance.

8.1. Research and academia

- ☆ HPLC is extensively used in research and academia for various applications.
- ☆ It supports studies in fields such as chemistry, biochemistry, pharmacology, environmental sciences, and material sciences.
- ☆ HPLC aids in the separation, purification, and characterization of compounds, enabling researchers to explore and understand complex systems.
- ☆ HPLC plays a crucial role in the analysis and purification of biomolecules, such as proteins, peptides, nucleic acids, and carbohydrates. It is employed in protein purification, amino acid analysis, DNA sequencing, and characterization of biopharmaceuticals.

8.2. Pharmaceuticals

- ☆ HPLC is the method of choice for pharmaceutical analysis, including drug discovery, development, and quality control. It is employed in the analysis of drug compounds, impurities, degradation products, and formulation analysis.
- ☆ HPLC ensures the identification and quantification of active pharmaceutical ingredients (APIs) and aids in the determination of their purity, stability, and dissolution profiles.

8.3. Environmental analysis

- ☆ HPLC is widely used in environmental monitoring and analysis.
- ☆ It enables the detection and quantification of pollutants, pesticides, herbicides, and other toxic compounds in air, water, soil, and biological samples.
- ☆ HPLC helps in assessing the impact of industrial activities, agricultural practices, and other factors on the environment.

8.4. Food and beverage

- ☆ HPLC plays a crucial role in food and beverage analysis.
- ☆ It is employed for the determination of additives, preservatives, vitamins, amino acids, pesticides, mycotoxins, and other contaminants in food and beverage products.
- ☆ HPLC ensures the safety, quality, and regulatory compliance of food and beverage products.

8.5. Forensic sciences

- ☆ HPLC is used in forensic analysis for the identification and quantification of drugs, metabolites, and toxins in biological samples.
- ☆ It aids in toxicological screenings, drug testing, and post-mortem analysis.
- ☆ HPLC is also utilized in the analysis of trace evidence, such as fibres, paints, and gunshot residue.

8.6. Petrochemical and industrial analysis

- ☆ HPLC is utilized in the analysis of petrochemicals, polymers, and industrial chemicals.
- ☆ It aids in the identification and quantification of organic compounds, antioxidants, surfactants, and impurities in industrial products.
- ☆ HPLC ensures quality control and process optimization in the petrochemical and manufacturing industries.

8.7. Clinical diagnostics

- ☆ HPLC is employed in clinical laboratories for the analysis of drugs, metabolites, and biomarkers in biological fluids.
- ☆ It aids in therapeutic drug monitoring, detection of drugs of abuse, analysis of hormones, and assessment of disease markers.
- ☆ HPLC enables accurate and sensitive quantification of analytes in clinical samples.

11 Other Chromatography Methods

CHAPTER SUMMARY

- Other than the major chromatographic methods discussed earlier, there are additional chromatographic methods with unique capabilities for separating and analyzing complex mixtures.
- Chiral chromatography is employed to separate enantiomers, which are mirror-image isomers.
- Countercurrent chromatography employs a biphasic liquid system and relies on the partitioning of the target compound between the two phases.
- Electrochromatography combines principles of both chromatography and electrophoresis, utilizing an electric field to drive the separation.
- Flash chromatography is a rapid technique that utilizes a stationary phase packed into a column for efficient separation.
- Hydrophilic interaction chromatography exploits the polar interactions between the stationary phase and hydrophilic compounds.
- Hydroxyapatite chromatography separates biomolecules based on their affinity for hydroxyapatite resin.
- Multi-dimensional chromatography involves the use of multiple separation modes to achieve higher resolution.
- Pyrolysis gas chromatography involves the thermal decomposition of a sample followed by gas chromatographic separation of the products.
- Supercritical fluid chromatography uses supercritical fluids as the mobile phase, providing enhanced efficiency and selectivity.

1. Chiral Chromatography

- Chiral chromatography is a specialized form of chromatography used to separate enantiomers (mirror-image isomers of a molecule). Enantiomers have the same physical and chemical properties but differ in their biological activity, pharmacological properties, and toxicological effects. Chiral chromatography separates enantiomers by exploiting the differences in their interactions with a chiral stationary phase.

- ☆ The chiral stationary phase is typically a chiral compound, such as cyclodextrin or a chiral amino acid, immobilized onto a solid support. The sample is passed through the stationary phase under controlled conditions, and the enantiomers are separated based on their differential interactions with the stationary phase.
- ☆ Chiral chromatography is widely used in the pharmaceutical and biotech industries for the separation and analysis of chiral compounds, such as drug molecules and natural products. It is also used in the food and cosmetic industries for the analysis of chiral additives and flavouring agents.
- ☆ One of the advantages of chiral chromatography is its high selectivity and resolution, which allows for the separation of closely related enantiomers. It is also a versatile technique, with different types of chiral stationary phases available to suit a range of separation requirements.
- ☆ Despite its advantages, chiral chromatography also has some limitations, such as its sensitivity to sample impurities and matrix effects, as well as high cost and technical complexity. Nonetheless, chiral chromatography remains a valuable technique for the separation and analysis of chiral compounds in a wide range of applications.

2. Countercurrent Chromatography

- ☆ Countercurrent chromatography (CCC) is a type of liquid-liquid chromatography that uses a biphasic solvent system to separate molecules. It is a form of liquid-liquid extraction that is based on the differential partitioning of molecules between two immiscible liquid phases.
- ☆ In CCC, the sample is injected into a column that contains two immiscible liquid phases, typically an aqueous phase and an organic phase. The column is then rotated to form a coiled column, which allows the two liquid phases to mix and separate in a countercurrent manner. As the sample moves through the column, it partitions between the two phases based on its solubility and polarity, resulting in the separation of the target molecules.
- ☆ CCC has several advantages over other chromatographic techniques. It is a highly efficient and selective technique, with the ability to separate complex mixtures with high resolution. It is also a gentle technique that does not require high pressure or high temperatures, which reduces the risk of sample degradation. CCC is also versatile, with the ability to separate molecules of varying sizes, polarities, and solubilities.
- ☆ Despite its advantages, CCC also has some limitations, such as the requirement for specialized equipment and expertise, as well as the need for large quantities of solvents. Nonetheless, CCC remains a powerful

technique for the separation and purification of a wide range of molecules, including natural products, peptides, and proteins.

3. Electrochromatography

- ☆ Electrochromatography (EC) is a hybrid technique that combines the principles of electrophoresis and chromatography.
- ☆ In EC, an electric field is applied to a chromatographic column, which results in the movement of charged molecules towards the oppositely charged electrode, allowing for the separation of the analytes.
- ☆ In EC, the stationary phase is usually a charged stationary phase that is immobilized on a support material, such as silica or polymer beads. The mobile phase is typically an aqueous buffer solution that contains the sample to be analyzed. When an electric field is applied to the column, the charged molecules in the sample move towards the oppositely charged electrode, resulting in the separation of the analytes based on their charge, size, and shape.
- ☆ EC has several advantages over other chromatographic techniques. It is a fast and efficient technique that allows for the separation of complex mixtures with high resolution. It is also a versatile technique that can be used to separate a wide range of charged molecules, including proteins, peptides, nucleic acids, and small molecules.
- ☆ Despite its advantages, EC also has some limitations, such as the requirement for specialized equipment and expertise, as well as the need for careful optimization of the operating conditions. Nonetheless, EC remains a valuable technique for the separation and analysis of charged molecules in a wide range of applications, including proteomics, genomics, and drug discovery.

4. Flash Chromatography

- ☆ Flash chromatography is a type of preparative chromatography that is commonly used for the separation and purification of organic compounds. It is a fast and efficient technique that uses a column packed with a stationary phase, typically silica gel or reversed-phase C18, and a mobile phase, which is usually a mixture of organic solvents and a polar modifier.
- ☆ In flash chromatography, the sample is loaded onto the top of the column and eluted with the mobile phase. The mobile phase is typically delivered by a pump at a flow rate of several millilitres per minute. As the sample passes through the column, the target compounds are retained on the stationary phase and separated from other components of the mixture.

- Flash chromatography is a cost-effective technique that allows for the separation of milligram to gram quantities of compounds in a short period of time. It is widely used in the pharmaceutical industry, as well as in research and development laboratories, for the purification of synthetic and natural products, peptides, and proteins.
- The main advantages of flash chromatography are its speed, simplicity, and versatility. However, it also has some limitations, such as the limited resolution and selectivity compared to other chromatographic techniques and the potential for peak tailing or band broadening. Nonetheless, flash chromatography remains a valuable technique for the separation and purification of organic compounds in a wide range of applications.

5. Hydrophilic Interaction Chromatography

- Hydrophilic interaction chromatography (HILIC) is a type of liquid chromatography that uses a hydrophilic stationary phase and a mobile phase that contains a high proportion of organic solvent and a low concentration of salts. HILIC is used primarily for the separation of polar and hydrophilic compounds, such as small organic molecules, peptides, and glycoproteins.
- In HILIC, the stationary phase is typically a polar material, such as silica, zirconia, or amino-bonded silica, that has a strong affinity for polar and hydrophilic analytes. The mobile phase contains a high percentage of organic solvents, such as acetonitrile or methanol, which helps to weaken the interactions between the analytes and the stationary phase.
- One of the advantages of HILIC is that it can provide high selectivity and resolution for polar and hydrophilic compounds that are difficult to separate by other chromatographic techniques, such as reversed-phase chromatography. HILIC also allows for the separation of isomers, positional isomers, and stereoisomers of polar compounds.
- HILIC has applications in a wide range of fields, including pharmaceuticals, metabolomics, proteomics, and bioanalysis. It is used for the separation and quantification of small molecules, such as amino acids, nucleotides, and organic acids, as well as larger molecules, such as peptides and glycoproteins. HILIC can also be used for the separation of chiral compounds by coupling with chiral stationary phases.

6. Hydroxyapatite Chromatography

- Hydroxyapatite chromatography is widely used in the field of biochemistry and biotechnology. It is based on the unique affinity of hydroxyapatite, a calcium phosphate mineral, for biomolecules such as proteins and nucleic acids. This chromatographic method offers several advantages, making it an essential tool in various research and industrial applications.

- ☆ The principle behind hydroxyapatite chromatography lies in the selective adsorption of target molecules onto the hydroxyapatite matrix. The binding occurs through a combination of electrostatic interactions, hydrogen bonding, and hydrophobic interactions. The specificity of the interaction is determined by the charge and conformation of the biomolecules, as well as the pH and salt concentration of the mobile phase.
- ☆ One of the major advantages of hydroxyapatite chromatography is its ability to separate biomolecules based on their structural and chemical differences. It can effectively separate proteins with different surface charges, sizes, and affinities. Moreover, hydroxyapatite chromatography is known for its mild and gentle elution conditions, which help preserve the structural integrity and biological activity of the separated molecules.
- ☆ Hydroxyapatite chromatography finds extensive applications in the purification and isolation of various biomolecules. It is commonly used for the purification of proteins, including enzymes, antibodies, and recombinant proteins. Additionally, it plays a crucial role in separating and purifying nucleic acids, such as DNA and RNA, from complex biological samples.
- ☆ In the field of biotechnology, hydroxyapatite chromatography is employed in the downstream processing of biotherapeutics, including vaccines, growth factors, and hormones. It offers high-resolution separation and purification, ensuring the removal of impurities and contaminants, thus resulting in high-purity final products.

7. Multi-dimensional Chromatography

- ☆ Multidimensional chromatography combines two or more separation mechanisms to increase separation selectivity and improve resolution. The concept of multidimensional chromatography is based on the fact that different analytes have different physicochemical properties and interact differently with different stationary phases, making it possible to separate complex mixtures into their individual components.
- ☆ Multidimensional chromatography usually involves two or more chromatographic columns, each using a different separation mechanism, such as reversed phase, ion exchange, size exclusion, or affinity chromatography. The eluent from the first column is then directed to a second column, where a different separation mechanism is used to further separate the components. This process can be repeated for additional dimensions, each with its own specific separation mechanism, to achieve even greater separation selectivity.

- Multidimensional chromatography has a wide range of applications in various fields, such as pharmaceuticals, proteomics, metabolomics, and environmental analysis. It is particularly useful for the separation of complex mixtures, such as biological samples, where there are often many similar compounds present. Multidimensional chromatography can provide a more detailed and accurate analysis of complex samples, allowing for the identification and quantification of components that would be difficult to separate using a single-dimensional chromatography technique.

8. Pyrolysis Gas Chromatography

- Pyrolysis gas chromatography (Py-GC) is an analytical technique used to identify and analyze the chemical composition of complex organic compounds. It involves heating a sample to high temperatures in a closed chamber, causing it to break down into smaller, volatile fragments that can be analyzed by gas chromatography.
- The process involves heating a small amount of the sample, usually between 0.1 to 5 mg, to temperatures of 500 to 1400 °C. This results in the breakdown of the sample into volatile fragments, which are then carried by a carrier gas into a gas chromatograph, where they are separated based on their physical and chemical properties.
- Py-GC is particularly useful for the analysis of complex organic compounds such as polymers, oils, and other organic materials. It can be used to identify the individual components of a sample and to determine their relative concentrations. Py-GC is also useful for analyzing the thermal stability of materials and can provide information on the chemical structure of the sample.
- Py-GC is widely used in various industries, including food and beverage, petrochemicals, polymers, and environmental analysis. It can be used to analyze a wide range of samples, including plastics, rubber, fibres, coatings, and adhesives. The technique provides a powerful tool for the identification and characterization of organic compounds, allowing for a better understanding of their properties and behaviour.

9. Supercritical Fluid Chromatography

- Supercritical fluid chromatography (SFC) is a type of chromatography that utilizes supercritical fluids as the mobile phase. Supercritical fluids are gases that are compressed and heated to a temperature and pressure above their critical point, where they exhibit both gas-like and liquid-like properties. The most commonly used supercritical fluid for SFC is carbon dioxide (CO_2).

- ☆ SFC is often used for the separation and analysis of non-volatile and thermally labile compounds such as natural products, pharmaceuticals, and polymers. SFC offers several advantages over traditional liquid chromatography and gas chromatography techniques, including faster separations, higher efficiency, and reduced use of organic solvents.
- ☆ In SFC, the supercritical fluid is pumped through a column containing the stationary phase, which can be either polar or nonpolar, depending on the sample being analyzed. The sample is dissolved or suspended in the mobile phase and is then introduced into the column. As the sample passes through the column, it interacts with the stationary phase, causing the separation of individual components.
- ☆ SFC has become an important analytical tool in many industries, including pharmaceuticals, natural products, and food and beverage. It offers a more environmentally friendly and efficient alternative to traditional chromatography techniques, making it an increasingly popular method of analysis.

12 Centrifugation Methods

CHAPTER SUMMARY

- Centrifugation is a powerful separation method used in various scientific and industrial fields.
- It involves the application of centrifugal force to separate components based on their sedimentation rates.
- Centrifugation methods include differential centrifugation, density gradient centrifugation, and ultracentrifugation, each offering unique ways to separate particles based on their properties and sedimentation rates.
- Centrifuge design and operation involve rotor types, speed and time parameters, temperature control, and safety measures to ensure efficient and safe centrifugation processes.
- Sample preparation involves the selection of appropriate sample containers, proper handling and labelling of samples, and ensuring the compatibility of samples with centrifugation conditions.
- Centrifugation has applications in various fields such as biology, biochemistry, pharmaceuticals, environmental sciences, and research, enabling separation, purification, and analysis of biomolecules, particles, and cellular components.
- Advances in centrifugation methods have led to the development of high-speed and ultracentrifuges, improved rotor designs, and advanced analytical methods, enabling more precise and efficient separations in various scientific and industrial applications.

1. Introduction

- Centrifugation is a widely used separation method that utilizes the principle of sedimentation and centrifugal force to separate particles or components of a mixture based on their density and size.
- It has applications in various fields, including biology, chemistry, medicine, and environmental sciences.

☆ Centrifugation works on the principle that when a mixture is subjected to high-speed rotation in a centrifuge, the denser particles or components move towards the bottom of the tube due to the action of centrifugal force. This separation is achieved by creating a density gradient within the sample, where the denser components sediment to the bottom while the lighter components remain at the top.

☆ Centrifugation methods can be classified into different types based on speed, rotor design, and separation mechanism. These include differential centrifugation, density gradient centrifugation, and ultracentrifugation. Differential centrifugation is used for the separation of particles with large size differences, while density gradient centrifugation is suitable for separating particles with similar sizes but different densities. Ultracentrifugation, which involves high-speed rotation, is used for separating macromolecules and particles with high precision.

2. Historical Background

☆ The concept of centrifugation dates back to the 19th century when Swedish scientist Carl Gustaf Patrik de Laval invented the first continuous centrifuge.

☆ In the 1920s, Theodor Svedberg developed the analytical ultracentrifuge, which enabled the determination of particle sizes and molecular weights based on sedimentation velocity. This breakthrough laid the foundation for modern ultracentrifugation methods.

☆ In the mid-20th century, centrifugation gained significant attention and became an indispensable method in various scientific disciplines.

☆ Over the years, advancements in centrifuge design, rotor technology, and control systems have further refined and expanded the applications of centrifugation in various scientific disciplines such as biochemistry, molecular biology, pharmaceuticals, and biotechnology.

☆ Modern centrifuges offer precise control over speed, acceleration, and deceleration rates, allowing for enhanced separation efficiency and sample handling.

3. Basic Principles of Centrifugation

☆ Centrifugation operates on the principle of centrifugal force, which is the force exerted on objects moving in a circular path.

☆ Understanding the principles of centrifugation is essential for performing effective separations.

3.1. Centrifugal force

☆ Centrifugal force is an apparent force that acts outwardly on objects rotating in a circular motion (Figure 1).

- When a sample is placed in a centrifuge and spun at high speeds, the centrifugal force causes the denser components to move outward and settle at the bottom of the tube or container. This separation occurs due to the difference in density between the components.
- Centrifugal force arises from the inertia of objects moving in a curved path. When an object rotates in a circle, it experiences an outward force that appears to push it away from the centre of rotation. This apparent force is known as centrifugal force and is mathematically represented as:

$$F = m \times \omega^2 \times r$$

where: F is the centrifugal force, m is the mass of the object, ω is the angular velocity (measured in radians per second) r is the distance from the object to the axis of rotation.

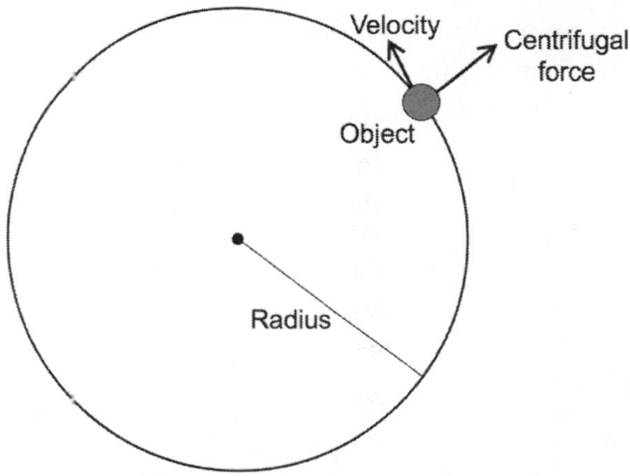

Figure 1: Schematic diagram depicting centrifugal force in a rotating system. As the object rotates at high velocity, the centrifugal force increases, causing the separation of components.

- The centrifugal force is directly proportional to the mass of the object, the square of the angular velocity, and the distance from the axis of rotation. As these parameters increase, the centrifugal force becomes stronger, resulting in greater separation or sedimentation effects in centrifugation.
- In centrifuge operations, various calculations are used to determine the necessary parameters for achieving specific separation goals. These calculations involve factors such as rotor speed, radius, and relative centrifugal force (RCF).
- By adjusting the rotor speed and radius, the RCF can be controlled to optimize the centrifugation process for different applications and sample types.

3.2. Sedimentation

- Sedimentation is the process by which particles or components in a sample separate and settle under the influence of gravity or centrifugal force.
- The rate of sedimentation depends on the size, shape, and density of the particles. Larger and denser particles sediment more rapidly, while smaller and less dense particles settle more slowly.
- Sedimentation can be used to separate solid particles from liquids or to separate different components within a mixture.
- The sedimentation process can be mathematically described using Stokes' law, which relates the settling velocity of a particle to its size and the properties of the medium.
- Stokes' law states that the settling velocity (v) of a spherical particle in a viscous medium is directly proportional to the square of the particle's radius (r) and the density difference (Δρ) between the particle and the medium and inversely proportional to the viscosity (η) of the medium.
- The formula for the settling velocity (v) according to Stokes' law is given by:

$$v=(2/9) \times g \times r^2 \times \Delta\rho/n$$

where: v is the settling velocity, g is the acceleration due to gravity, r is the particle radius, Δρ is the density difference between the particle and medium, η is the viscosity of the medium.

- Sedimentation is an important process in various fields, including environmental science, wastewater treatment, pharmaceuticals, and geological studies. By understanding the principles and employing appropriate formulae and calculations, scientists and engineers can effectively control and optimize the sedimentation process for efficient separation and analysis of particles.

3.2.1. Sedimentation coefficient

- The sedimentation coefficient is a crucial parameter in sedimentation and ultracentrifugation.
- It quantifies the rate at which particles or molecules move through a liquid under the influence of centrifugal force. It is measured in Svedberg units (S) and has a unit of second.
- The sedimentation coefficient depends on the size, shape, and density of the particles or molecules. It provides valuable insights into their molecular weight, shape, and structure.
- The Svedberg unit is a convenient way to express the sedimentation coefficient, with larger values indicating slower sedimentation rates.

- ☆ The sedimentation coefficient is widely used in biophysical and biochemical research, helping to characterize macromolecules, study protein-protein interactions, determine molecular assemblies, and investigate the kinetics of molecular processes.

3.2.2. Svedberg unit (S)

- ☆ The Svedberg unit (S) is a measure of the sedimentation coefficient or rate used in ultracentrifugation to determine the size and shape of macromolecules. It is named after Theodor Svedberg.
- ☆ The Svedberg unit is calculated using the equation:

$$S = (t - t_0) / \omega^2$$

  where: t is the time of sedimentation, t_0 is the reference time for solvent, and ω is the angular velocity of the rotor.
- ☆ The Svedberg unit is a non-linear scale that represents the sedimentation coefficient, with larger values indicating slower-settling particles. It is widely used in the field of biochemistry to characterize the size and structure of biological macromolecules, such as proteins and nucleic acids, providing valuable insights into their biological functions and interactions.

Note

- ☆ A sedimentation coefficient of 1×10^{-13} second is defined as one Svedberg, S. ($1S = 10^{-13}$ sec)
- ☆ The sedimentation coefficient of most biological particles is very small. An rRNA molecule having a sedimentation coefficient of 5×10^{-13} is said to have a value of 5s rRNA.
- ☆ The larger the particle, the higher the Svedberg's unit and the faster its sedimentation rate.
- ☆ Thus, as the molecular weight of the particle increases, the S increases.
- ☆ The approximate sedimentation rates (in Svedberg units, S) of some common biological particles are: Ribosomes: 70S in prokaryotes, 80S in eukaryotes; Small proteins: 2S to 5S; Large proteins: 10S to 20S; Viruses: 10S to 400S; Mitochondria: 70S to 80S; Microsomes: 80S to 120S; Nuclei: 100S to 200S; Protein complexes: 10S to 30S.
- ☆ It is important to note that these sedimentation rates are approximate values and can vary depending on the specific conditions of the centrifugation experiment, such as rotor type, speed, and temperature.

3.3. Centrifugation speed and time

- ☆ The speed at which a centrifuge rotates, commonly measured in revolutions per minute (RPM) or relative centrifugal force (RCF), is a crucial parameter in centrifugation.
- ☆ Higher speeds generate greater centrifugal forces, leading to faster and more efficient separations.
- ☆ The duration of centrifugation, or the time required for particles to sediment, depends on the sample and the desired separation. Longer centrifugation times allow for better separation of components.
- ☆ RPM and RCF are important parameters in centrifugation.
- ☆ RPM refers to the number of rotations the centrifuge rotor makes per minute, while RCF is a measure of the effective gravitational force experienced by the sample during centrifugation.
- ☆ RCF is calculated using the equation:

$$RCF = \left(1.12 \times 10^{-5}\right) \times (r) \times (N)^2$$

where: r is the radius of the centrifuge rotor (in centimetres), and N is the rotor speed (in revolutions per minute, RPM).

- ☆ RPM and RCF play crucial roles in determining the sedimentation rate, pelleting efficiency, and separation capabilities of a centrifuge.
- ☆ The relationship between relative centrifugal force (RCF) and the acceleration due to gravity (g) can be described as follows: RCF is directly proportional to g. RCF is a measure of the gravitational force acting on a sample in a centrifuge, and it determines the sedimentation rate and separation efficiency. The value of g depends on the geographical location, and it is typically around 9.8 m/s^2 on the Earth's surface.
- ☆ The RCF generated by a rotor depends on the speed of the rotor in revolutions per minute (rpm) and the radius of rotation (i.e., the distance from the axis of rotation).
- ☆ The equation to calculate the RCF from a known rpm and radius of rotation is:

$$RCF = \frac{\omega^2 r}{g}$$

where: r is the radius of the centrifuge rotor (in centimetres), g is the acceleration due to gravity (cm/sec^2), and ω is the angular velocity.

- ☆ The angular velocity can be calculated using:

$$\omega = \frac{2\pi \ x \ rpm}{60}$$

- ☆ By controlling the RPM and RCF, scientists can optimize the centrifugation conditions for various applications, such as sample preparation, isolation of specific components, or separation of particles based on size and density.

3.4. Centrifuge design

- ☆ Centrifuges are designed with various features to optimize separation efficiency. The rotor design is an essential consideration, as it determines the orientation of the sample during centrifugation.
- ☆ Fixed-angle rotors hold the sample at a fixed angle relative to the axis of rotation, allowing efficient separation. Swinging-bucket rotors allow samples to swing out during centrifugation, facilitating better separation of large volumes or fragile samples.

3.5. Sample preparation

- ☆ Proper sample preparation is crucial for successful centrifugation.
- ☆ The sample should be carefully collected and prepared to ensure accurate and reproducible results. It may involve dilution, filtration, or extraction of specific components.
- ☆ The selection of appropriate sample containers, such as centrifuge tubes or plates, is also essential to ensure compatibility with the centrifuge and facilitate efficient separations.

4. Types of Centrifugation Methods

- ☆ There are three broad types of centrifugation methods: differential centrifugation, density gradient centrifugation, and ultracentrifugation.
- ☆ Differential centrifugation separates components based on their sedimentation rates, density gradient centrifugation utilizes density gradients to separate particles, and ultracentrifugation employs high speeds and forces for precise separation.
- ☆ Each type of centrifugation has its own advantages and applications, enabling researchers to isolate and study specific components of interest.
- ☆ These methods have revolutionized fields such as cell biology, biochemistry, and molecular biology, facilitating advancements in research, diagnostics, and drug development.

4.1. Differential centrifugation

- ☆ Differential centrifugation is a method used to separate cellular components based on their size and density (Figure 2).
- ☆ It involves a series of centrifugation steps to isolate specific organelles and subcellular structures.

☆ Differential centrifugation can be of two types, preparative and analytical differential centrifugation. Preparative differential centrifugation is used to isolate specific components for further analysis or purification, while analytical differential centrifugation is employed to study the distribution and composition of cellular components.

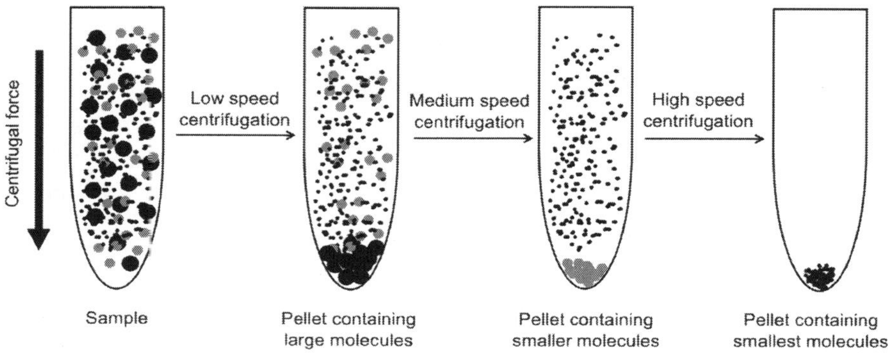

Figure 2: Differential centrifugation process. The diagram illustrates the sequential steps involved in differential centrifugation: low-speed centrifugation, supernatant removal, and subsequent high-speed centrifugation to obtain fractions with varying sedimentation rates. The separation of particles is based on their sedimentation rates.

4.1.1. Preparative differential centrifugation

☆ Preparative differential centrifugation is a powerful method used in biological and biochemical research for the separation and isolation of cellular components based on their size and density. This method enables the extraction of specific organelles and subcellular structures for further analysis and study.

☆ During preparative differential centrifugation, a sample containing cells or tissue homogenate is subjected to a series of centrifugation steps at gradually increasing speeds. The goal is to separate the various cellular components based on sedimentation rates. As the centrifuge rotor spins, particles in the sample experience a centrifugal force that causes them to sediment towards the bottom of the tube.

☆ The process starts with a low-speed centrifugation step, which removes large debris and intact cells. The supernatant, containing smaller particles such as nuclei and mitochondria, is then collected and subjected to a higher-speed centrifugation step. This step allows for the separation of these organelles from the cytosol and other cellular components.

☆ Subsequent rounds of centrifugation at increasing speeds help to isolate different organelles with higher specificity. By carefully adjusting the centrifugation conditions, researchers can obtain fractions enriched with specific cellular components of interest.

4.1.2. Analytical differential centrifugation

- ☆ Analytical differential centrifugation is a method used to analyse the distribution and properties of particles in a sample. Unlike preparative differential centrifugation, which aims to isolate specific components, analytical differential centrifugation focuses on obtaining information about the relative abundance and sedimentation behaviour of particles.
- ☆ In analytical differential centrifugation, a sample is subjected to a series of centrifugation steps at different speeds, similar to the preparative approach. However, instead of isolating distinct fractions, the goal is to obtain sedimentation profiles that reveal the distribution of particles based on their sedimentation rates.
- ☆ By collecting and analysing the supernatant and pellet fractions obtained after each centrifugation step, researchers can determine the sedimentation behaviour of different particles in the sample. This information can provide insights into the size, density, and composition of the particles.
- ☆ Analytical differential centrifugation is commonly used in proteomics and molecular biology studies to investigate the subcellular localization of proteins and the association of proteins with cellular structures. By comparing the sedimentation profiles of different samples, researchers can identify changes in protein localization or interactions under different experimental conditions.

4.2. Density gradient centrifugation

- ☆ Density gradient centrifugation is a powerful method widely used in various scientific disciplines to separate particles and molecules based on density.
- ☆ This method involves the creation of a density gradient within a centrifuge tube, allowing particles to migrate and form distinct bands or layers depending on their buoyant density.
- ☆ Density gradient centrifugation offers several advantages in separation methods. It provides a high degree of resolution and allows for the separation of particles with close densities or similar sizes. It is a versatile method that can be applied to various sample types and has applications in fields such as molecular biology, biochemistry, and virology. Researchers utilize density gradient centrifugation to isolate specific components, purify samples, study molecular interactions, and analyse biological macromolecules.
- ☆ Two common types of density gradient centrifugation are rate-zonal centrifugation and isopycnic centrifugation. Each method offers unique advantages and applications in different fields of research.

4.2.1. Rate-zonal centrifugation

- ☆ Rate-zonal centrifugation separates particles based on their size and shape.
- ☆ The method involves the use of a density gradient medium, such as sucrose, which creates a density gradient in the centrifuge tube.
- ☆ The principle behind rate-zonal centrifugation is that during centrifugation, particles sediment at different speeds according to their size and shape.
- ☆ A sample containing particles of varying sizes is layered onto a density gradient medium. As the centrifuge rotor spins, the particles in the sample move through the density gradient, with larger particles sedimenting faster than smaller ones (Figure 3).
- ☆ The differential sedimentation rates lead to the formation of distinct bands or zones in the centrifuge tube, each representing particles of a specific size range. The separation is achieved by carefully controlling the centrifugal force and duration of centrifugation to ensure that the particles of interest sediment to their respective positions within the tube.
- ☆ After centrifugation, the tube is carefully fractionated, and each zone is collected separately.
- ☆ Rate-zonal centrifugation is commonly used to separate proteins, nucleic acids, and subcellular organelles based on sedimentation rates.

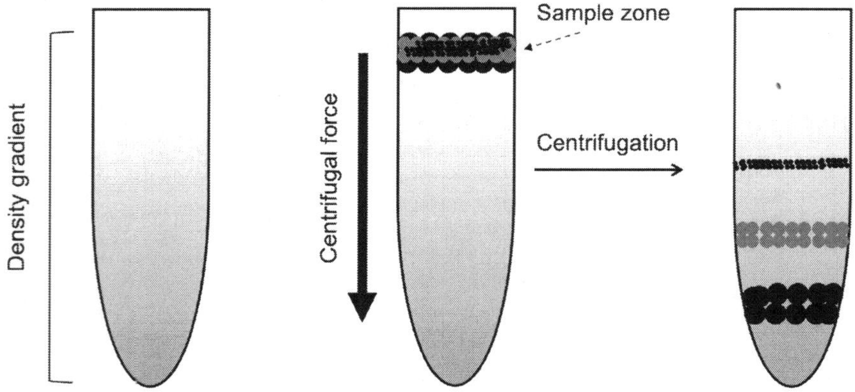

Figure 3: Rate-zonal centrifugation. The separation of particles is based on their sedimentation rates. The heavier particles settle closer to the bottom, while lighter particles migrate to low-density regions, forming distinct bands.

4.2.2. Isopycnic centrifugation

- ☆ Isopycnic centrifugation, also known as equilibrium centrifugation or buoyant density centrifugation, separates particles based on their buoyant density.

- ☆ The principle of isopycnic centrifugation is based on the fact that in a density gradient, particles will move through the gradient until they reach a point where their buoyant density equals the density of the medium. This equilibrium position is determined by the particle's density and shape.
- ☆ The process of isopycnic centrifugation involves the preparation of a density gradient using a gradient medium, such as caesium chloride. The sample containing the particles of interest is then loaded on top of the gradient and centrifuged.
- ☆ During centrifugation, particles migrate through the gradient until they reach a position where their density matches the density of the surrounding medium. The equilibrium position depends on the density of the particle, with denser particles settling at lower positions in the gradient and less dense particles settling at higher positions.
- ☆ Once the centrifugation is complete, the tube is carefully fractionated, collecting fractions containing particles that are sedimented at specific positions in the gradient. The separated particles can then be further analysed or purified for downstream applications.
- ☆ Isopycnic centrifugation is useful for separating particles with similar densities, such as DNA fragments, RNA, and viruses. It allows for the purification and concentration of specific particles based on their buoyant density.

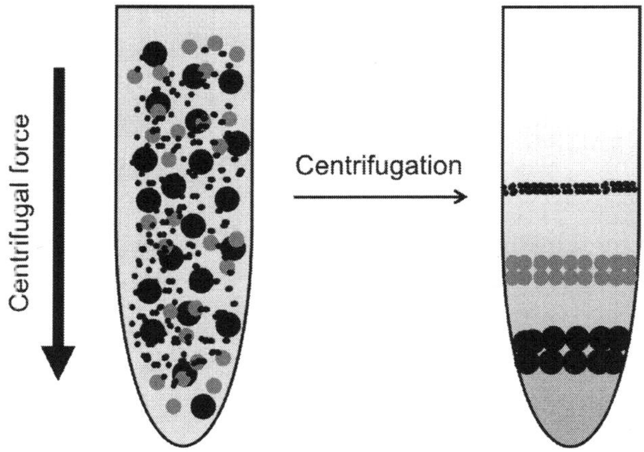

Figure 4: Isopycnic centrifugation. The separation of particles is based on their equilibrium positions in a density gradient. The denser particles settle closer to the bottom, while lighter particles migrate to low-density regions, forming distinct bands.

Note

- ☆ Rate-zonal and isopycnic centrifugation are two distinct methods used for biomolecular separation. While both methods involve centrifugal force and density gradients, they differ in their principles and applications.
- ☆ Rate-zonal centrifugation is based on the differential sedimentation rates of particles in a centrifugal field. The separation is achieved by layering a sample onto a density gradient medium and as the centrifuge rotor spins, the particles sediment at different speeds based on their size and shape. This results in the formation of distinct bands or zones within the tube, with each zone representing particles of a specific size range. Rate-zonal centrifugation is commonly used for size-based separation of macromolecules, such as proteins and nucleic acids.
- ☆ On the other hand, isopycnic centrifugation relies on the differential buoyant densities of particles in a density gradient medium. Here, the sample is layered onto a pre-formed density gradient, and as the centrifuge rotor spins, particles migrate within the gradient until they reach their isopycnic point, where their density matches that of the surrounding medium. This results in the formation of distinct bands or zones, with each zone representing particles of the same density. Isopycnic centrifugation is commonly used for the purification and separation of particles with similar densities, such as viruses, subcellular organelles, or biomolecules with bound ligands.

4.3. Ultracentrifugation

- ☆ Ultracentrifugation is a powerful method that utilizes high centrifugal forces to separate and analyse particles based on sedimentation rates.
- ☆ It has revolutionized the field of molecular biology and biochemistry by enabling the study of macromolecules and their interactions.
- ☆ There are two main types of ultracentrifugation: analytical ultracentrifugation and preparative ultracentrifugation. Each method has its own specific applications and benefits.

4.3.1. Analytical ultracentrifugation

- ☆ Analytical ultracentrifugation (AUC) is a powerful method used in the field of biophysics and biochemistry to study the properties and behaviour of macromolecules in solution.
- ☆ It provides valuable insights into the size, shape, molecular weight, and interactions of biological molecules, such as proteins, nucleic acids, and macromolecular complexes.
- ☆ The principle behind AUC is the separation of particles based on their sedimentation rates under high centrifugal forces. A sample is loaded into a specialized rotor and subjected to high speeds, typically ranging from 40,000 to 150,000 revolutions per minute (rpm), in an ultracentrifuge. As the rotor spins, particles sediment towards the bottom of the cell based on their mass and shape.
- ☆ The detection system in an AUC plays a crucial role in analysing and monitoring the sedimentation process of particles. It typically consists of an optical detection system, which allows for real-time measurement of the concentration or absorbance of the sample as it undergoes centrifugation. This system utilizes an ultraviolet (UV) or visible light source, a detection cell, and a photodetector. As the particles sediment, changes in the sample's absorbance or concentration are detected, providing valuable information about the sample's composition and sedimentation behaviour. The data on the sedimentation profile over time can be used to determine important parameters of the molecules, including their molecular weight, shape, and association/dissociation kinetics.

4.3.1.1. Basic instrumentation of an AUC

- ☆ An AUC typically consists of the following components (Figure 5):

1. **Rotor:** The rotor is a crucial part of the AUC. It holds the sample cells or tubes containing the sample solution.

2. **Sample cells:** These are tubes or cells that hold the sample solution. They are placed in the rotor and are carefully balanced to maintain equilibrium during centrifugation.
3. **Detection system:** The detection system measures the sedimentation or migration of the sample components during centrifugation. It typically includes an optical system, such as a laser beam or UV-visible light source, and a detector, such as a photomultiplier tube (PMT) or a camera.
4. **Centrifuge chamber:** The centrifuge chamber houses the rotor and provides a controlled environment for the centrifugation process. It includes temperature control systems to maintain a constant temperature throughout the experiment.
5. **Drive system:** It generates the rotational force required to spin the rotor at high speeds. It consists of a motor, a drive shaft, and associated control mechanisms.
6. **Data acquisition and analysis system:** This system collects the data from the detection system and processes it to obtain sedimentation or migration profiles of the sample components. It includes software for data analysis and visualization.

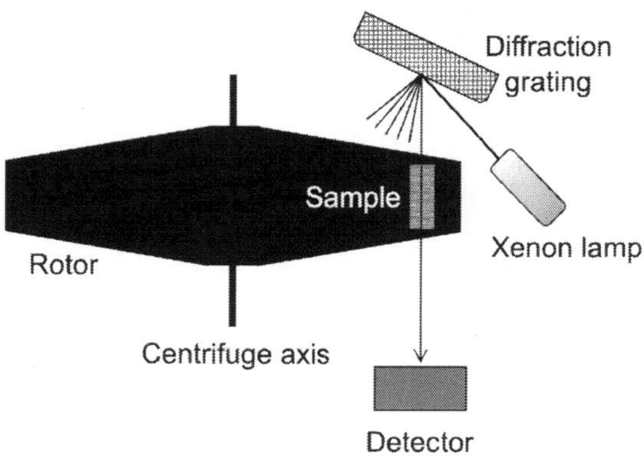

Figure 5: Construction of a typical AUC. This figure depicts a cross-section of a typical AUC. The AUC consists of a rotor, sample cells, and detection systems. The rotor spins at high speeds, generating a centrifugal force that causes particles in the sample to sediment. The detection systems measure the sedimentation of particles, providing information about their size, shape, and molecular interactions.

4.3.1.2. Types of AUC

☆ There are two main types of AUC methods: sedimentation velocity (SV) and sedimentation equilibrium (SE).

1. **Sedimentation velocity (SV):** SV AUC is commonly used to determine the size and shape of macromolecules in solution. During centrifugation, the particles migrate through the sample cell at different velocities based on size. By analysing the sedimentation profiles, scientists can obtain information about the molecular weight, shape, and conformational changes of the macromolecules. SV AUC is particularly useful for studying the aggregation, self-association, and oligomeric states of proteins and nucleic acids.
2. **Sedimentation equilibrium (SE):** SE AUC is employed to measure the distribution of macromolecules between different oligomeric states or to determine their association constants. In SE AUC, the centrifugation is performed at a lower speed, allowing the system to reach equilibrium. By measuring the concentration gradient of the macromolecule along the length of the cell, researchers can calculate the relative amounts of each oligomeric species present in the sample. SE AUC is valuable for studying protein-protein interactions, macromolecular assembly, and equilibrium constants.

- ☆ AUC offers several advantages over other biophysical methods. Firstly, it is a label-free method that does not require chemical modifications or immobilization of the molecules being studied. This allows for the analysis of macromolecules in their native, physiologically relevant states without any potential artefacts introduced by labelling. Secondly, AUC provides high-resolution data, allowing researchers to obtain detailed information about the size, shape, and interactions of macromolecules. Additionally, AUC can work with complex mixtures, making it suitable for studying multi-component systems.
- ☆ Applications of AUC are diverse. It is used in understanding protein folding, protein-protein interactions, and self-association of proteins and nucleic acids, as well as characterizing macromolecular complexes and viruses. AUC is used in drug discovery and development to evaluate protein-drug interactions, assess the quality of biopharmaceuticals, and study protein stability and aggregation. It also plays a significant role in the field of structural biology, assisting in the determination of macromolecular structures and providing data for computational modelling.

4.3.2. Preparative ultracentrifugation

- ☆ Preparative ultracentrifugation (PUC) is used for the separation and purification of biomolecules based on their size, shape, density, or molecular weight.

- ☆ It involves the use of ultracentrifugation at high speeds to separate and collect specific components from a complex mixture.
- ☆ PUC is widely employed in various fields, including biochemistry, molecular biology, and biotechnology, for the isolation and enrichment of biomolecules for further analysis or downstream applications.
- ☆ The process of PUC involves several steps:

1. **Sample preparation:** The first step is to prepare the sample for centrifugation. This may involve cell lysis, homogenization, and removal of cellular debris or other unwanted components. The sample is typically prepared in a buffer or medium that maintains the stability and activity of the biomolecules of interest.
2. **Centrifugation conditions:** The next step is to set the appropriate centrifugation conditions. This includes selecting the rotor type, speed, and temperature. PUCs are equipped with rotors capable of holding larger volumes and accommodating specialized sample holders or tubes designed for sample fractionation.
3. **Loading and centrifugation:** The sample is loaded into the appropriate centrifuge tubes or containers and placed in the rotor. The rotor is then placed in the ultracentrifuge, and the centrifugation is carried out at high speeds. The centrifugal force causes the biomolecules to sediment based on their physical properties, such as size, shape, or density.
4. **Fractionation and collection:** During the centrifugation process, the biomolecules of interest separate into distinct bands or zones along the length of the tube. These zones represent different components or fractions. Once the centrifugation is complete, the fractions are carefully collected using fractionation methods such as gradient fractionation, fractionation by density, or fractionation based on sedimentation rate.

- ☆ PUC offers several advantages for biomolecule separation and purification. It allows for the isolation of specific components from complex mixtures, enabling the enrichment of target molecules for further analysis or downstream applications. The method provides high-resolution separation, preserving the integrity and activity of the separated biomolecules.
- ☆ Applications of PUC are vast and diverse. It is commonly used in the purification of proteins for biochemical and structural studies. The method allows for the separation of protein isoforms, purification of protein complexes, and removal of contaminants. PUC is also employed in the isolation of nucleic acids, such as DNA and RNA, for various molecular biology applications.

☆ Moreover, PUC plays a crucial role in virology, enabling the purification and concentration of viruses. It is essential for vaccine development and the study of viral structure and assembly. Furthermore, the method is employed in the separation and purification of subcellular organelles and membrane fractions, facilitating the investigation of cellular compartments and their functions.

5. Centrifuge design and operation

☆ Centrifuges are essential laboratory instruments used for separating and sedimenting particles or components of a sample based on their density, size, or molecular weight.

☆ The design and operation of a centrifuge play a critical role in achieving efficient and accurate separations.

☆ Proper maintenance and calibration of centrifuges are also crucial for their safe and effective operation. Regular inspection, cleaning, and servicing of centrifuge components, including rotors and seals, ensure optimal performance and prevent mechanical failures.

☆ The key aspects of centrifuge design and operation are discussed below.

5.1. Rotor types and selection

☆ Centrifuge rotors are available in various designs to accommodate different sample types and separation requirements.

☆ **Fixed-angle rotors** hold the sample at a fixed angle relative to the axis of rotation, providing efficient separation of particles.

☆ **Swinging-bucket rotors** allow samples to swing out during centrifugation, providing better separation of large volumes or fragile samples.

☆ Rotor selection depends on factors such as sample volume, density, and required separation efficiency.

5.2. Speed and time parameters

☆ Centrifuges operate at specific speeds or angular velocities, measured in revolutions per minute (RPM) or relative centrifugal force (RCF).

☆ The speed and time parameters are crucial for achieving the desired separation.

☆ The appropriate speed depends on the sample type and the forces required to sediment or separate the components effectively.

☆ Time parameters are determined based on the sedimentation properties of the particles or components being separated.

5.3. Temperature control

- Temperature control is important in many centrifugation applications, especially those involving sensitive samples.
- Centrifuges may have built-in temperature control mechanisms, such as refrigeration systems or water baths, to maintain the desired temperature during centrifugation.
- Temperature control helps preserve the stability and integrity of biological samples and prevents degradation or denaturation of sensitive molecules.

5.4. Safety considerations

- Centrifuges operate at high speeds, generating significant centrifugal forces. Safety considerations are paramount to protect the operator and prevent accidents.
- Centrifuges should have safety features such as lid-locking mechanisms, automatic rotor imbalance detection, and emergency stop buttons.
- It is essential to follow proper loading and unloading procedures, adhere to recommended maximum speed limits, and use appropriate protective equipment, such as gloves and goggles.

6. Sample preparation for centrifugation

- The success of centrifugation largely depends on proper sample preparation. This involves several considerations, including the choice of sample containers and sample preparation methods.

6.1. Choice of sample containers

- The selection of appropriate sample containers is crucial to ensure the integrity and stability of the sample during centrifugation.
- Factors such as the sample volume, type, and compatibility with the centrifuge rotor should be considered.
- Commonly used containers include centrifuge tubes, microcentrifuge tubes, and ultracentrifuge tubes made from materials like plastic or glass.

6.2. Sample preparation methods

- Proper sample preparation ensures that the sample is in an appropriate form for centrifugation and minimizes the chances of experimental artefacts.
- It allows for efficient separation, purification, and analysis of target components in the sample.
- Two commonly employed methods are cell lysis and homogenization and density gradient preparation.

1. **Cell lysis and homogenization:** Cell lysis involves breaking down cell membranes to release cellular components for further analysis. This can be achieved through mechanical methods such as sonication or chemical agents. Homogenization is used to disrupt tissues and ensure a uniform distribution of components in the sample. Methods like mortar and pestle, bead beating, or high-pressure homogenizers are commonly employed.
2. **Density gradient preparation:** Density gradient centrifugation is a method used for separating particles based on their buoyant density. To prepare a density gradient, a gradient-forming material, such as sucrose or iodixanol, is mixed with the sample. This creates a density gradient, and upon centrifugation, particles migrate and band at positions corresponding to their density.

7. Advancements and Innovations in Centrifugation Methods

- ☆ Centrifugation methods have evolved significantly over the years, driven by advancements in technology and the need for improved separation and analysis methods.
- ☆ These advances in centrifugation methods have revolutionized various fields, including biomedical research, drug discovery, molecular biology, and materials science. They have enabled faster, more precise separations, improved sample handling, and integration with other analytical methods, leading to enhanced data quality and deeper insights into complex biological and physical systems.

7.1. Microscale and nanoscale centrifugation

- ☆ Recent advancements have focused on miniaturizing centrifugation methods for microscale and nanoscale applications.
- ☆ Microcentrifuges and ultracentrifuges with smaller volumes and specialized rotors enable researchers to work with limited sample volumes, making them ideal for applications in genomics, proteomics, and nanotechnology.

7.2. High-speed centrifugation

- ☆ High-speed centrifugation methods have revolutionized the field, allowing for rapid separations and increased sample throughput.
- ☆ Advances in rotor design, materials, and balancing technologies have enabled researchers to achieve higher rotational speeds, resulting in improved separation efficiency and reduced processing times.

7.3. Centrifugation in microgravity environments

- ☆ Centrifugation in microgravity environments, such as space, offers unique opportunities for studying biological and physical phenomena.

☆ In these settings, gravitational forces are eliminated or significantly reduced, allowing for more precise control of sedimentation and separation processes.

☆ Microgravity centrifuges have been developed for space missions to study the effects of gravity on biological samples and to perform experiments in extraterrestrial environments.

7.4. Miniaturization and automation

☆ Miniaturization and automation have transformed centrifugation methods, enabling high-throughput analysis, improved precision, and reduced sample and reagent consumption.

☆ Microfluidic devices, lab-on-a-chip systems, and automated sample handling systems have made centrifugation more efficient, reproducible, and accessible for a wide range of applications.

7.5. Integration with other analytical methods

☆ Centrifugation methods have been integrated with other analytical methods to enhance sample preparation and analysis. For example, centrifugation coupled with mass spectrometry, spectroscopy, or microscopy enables the comprehensive characterization of complex samples and facilitates multidimensional analysis.

8. Applications

☆ Centrifugation methods have widespread applications in various fields, including biotechnology, clinical diagnostics, and industrial sectors. These methods enable the separation, purification, and analysis of different substances based on their physical properties.

8.1. Research and academia

☆ **Protein purification:** Centrifugation plays a crucial role in protein purification, allowing for the separation of proteins from complex mixtures. By utilizing methods such as differential centrifugation or density gradient centrifugation, specific proteins can be isolated and purified for further analysis, characterization, and use in research or industrial applications.

☆ **Cell fractionation:** Centrifugation enables the separation of cellular components, such as organelles, membranes, and cytoplasmic fractions. This method aids in studying cell structure, organelle function, and biochemical processes. It is essential for understanding cellular mechanisms and developing therapies for various diseases.

- ☆ **Nucleic acid extraction:** Centrifugation methods are employed for the extraction and purification of nucleic acids, including DNA and RNA. By subjecting samples to centrifugal forces, nucleic acids can be separated from other cellular components, facilitating genetic analysis, gene expression studies, and molecular diagnostics.
- ☆ **Virus and particle separation:** Centrifugation is crucial in isolating and purifying viruses and other particles. Using high-speed centrifugation, viruses can be separated from host cells, debris, and contaminants. This is vital for studying the viral structure, developing vaccines, and producing viral vectors for gene therapy.

8.2. Clinical and diagnostic applications

- ☆ **Blood cell separation:** Centrifugation is extensively used in clinical laboratories to separate blood cells from plasma or serum. This enables the analysis of blood components, such as red and white blood cells, platelets, and plasma proteins. It is vital for diagnosing various hematological disorders and monitoring patients' health.
- ☆ **Urine sediment analysis:** Centrifugation is employed to concentrate and separate cellular components and particles in urine samples. This aids in the examination of urinary sediments, identification of abnormal cells, crystals, or bacteria, and diagnosis of kidney diseases and urinary tract infections.
- ☆ **Cerebrospinal fluid examination:** Centrifugation is utilized in the analysis of cerebrospinal fluid (CSF), which surrounds the brain and spinal cord. By centrifuging CSF samples, cellular elements, infectious agents, and abnormal proteins can be separated and identified, aiding in the diagnosis of neurological disorders and infections.

8.3. Industrial applications

- ☆ **Chemical and pharmaceutical industry:** Centrifugation methods are employed for the separation, purification, and concentration of chemicals, pharmaceuticals, and bioproducts. It plays a crucial role in processes such as fermentation, cell culture, and downstream processing, enabling the isolation of desired compounds and removal of impurities.
- ☆ **Food and beverage industry:** Centrifugation is utilized in the food and beverage industry for the separation of solids, liquids, and suspensions. It aids in the clarification of juices, separation of fats and oils, concentration of flavours, and removal of impurities, ensuring product quality and purity.

- ☆ **Environmental applications:** Centrifugation plays a significant role in environmental applications, including wastewater treatment, soil analysis, and environmental monitoring. It allows for the separation of suspended particles, sediments, and contaminants, aiding in the purification of water, analysis of pollutants, and monitoring environmental impact.

13 Other Separation Methods

CHAPTER SUMMARY

- There are additional separation methods such as filtration, distillation, extraction, evaporation, crystallization, precipitation, electrolysis, and solvent extraction.
- Filtration is the separation of solids from liquids or gases through a porous medium.
- Distillation separates substances based on differences in their boiling points.
- Extraction transfers a desired substance from a mixture to a solvent.
- Evaporation removes a solvent by vaporizing it, leaving behind the solute.
- Crystallization induces the formation of crystals to separate a solute from a solvent.
- Precipitation forms insoluble particles by adding a precipitating agent to a solution.
- Electrolysis uses an electric current to drive a non-spontaneous chemical reaction and separate components.
- Solvent extraction transfers a solute from one solvent phase to another to separate mixtures.

1. Filtration

- ☆ Filtration involves the separation of solid particles from a liquid or gas mixture.
- ☆ It is a versatile process employed in various industries, laboratories, and everyday applications.
- ☆ By passing the mixture through a porous medium or filter, filtration allows the liquid or gas to pass through while retaining the solid particles.
- ☆ Filtration offers several advantages as a separation technique. It is a relatively straightforward and cost-effective method that can be easily implemented on various scales. Filtration allows for the efficient removal

of solid particles, facilitating the purification and clarification of liquids or gases. The separated particles can often be collected and further analyzed if needed. Additionally, filtration is a non-destructive process, as it does not alter the chemical composition or properties of the substances being separated.

☆ However, there are certain limitations to filtration. Fine particles or particles with sizes close to the pore size of the filter may clog the filter and reduce its efficiency. To overcome this, pre-filtration or multiple filtration steps may be required. Additionally, some particles may have electrostatic charges or adhere strongly to the filter surface, making them more challenging to separate. In such cases, additional treatments or alternative separation methods may be necessary.

1.1. Principle

☆ The fundamental principle behind filtration is the difference in size between the particles and the pores of the filter medium.

☆ The filter acts as a barrier, preventing the passage of particles that are larger than the pore size.

☆ The efficiency of filtration depends on several factors, including the pore size of the filter, the nature of the particles, and the operating conditions.

1.2. Types

☆ There are several types of filtration methods, each suited for different applications and particle sizes.

☆ The choice of filtration technique depends on factors such as the size of the particles, the desired filtration rate, the nature of the mixture, and the purpose of the separation.

☆ Understanding the various types of filtration methods allows for the selection of the most appropriate technique to achieve efficient separation and purification of substances.

1. **Gravity filtration:** This is the most common and straightforward form of filtration. It involves pouring the mixture into a funnel containing filter paper, and gravity pulls the liquid through while retaining the solid particles. Gravity filtration is suitable for large-scale applications where the particle size is relatively large, and a slow filtration rate is acceptable.

2. **Vacuum filtration:** Vacuum filtration enhances the efficiency of filtration by applying a vacuum to the system. This increases the filtration rate as the pressure difference helps draw the liquid through the filter medium. It is particularly useful when a faster separation is required or when dealing with fine particles.

3. **Pressure filtration:** In pressure filtration, pressure is applied to the mixture to force it through the filter medium. This method is especially effective when a high filtration rate is desired or when dealing with viscous liquids. Pressure filtration is commonly used in industrial applications and large-scale processes.
4. **Centrifugal filtration:** Centrifugal filtration takes advantage of centrifugal force to separate solid particles from the mixture. The mixture is placed in a rotating container, and the centrifugal force causes the solid particles to move towards the outer edge while the liquid remains near the centre. This method is particularly useful for separating fine particles or for separating a large volume of suspension.
5. **Membrane filtration:** Membrane filtration uses a semi-permeable membrane to separate particles based on size. The membrane contains tiny pores that allow the passage of specific-sized particles while retaining larger particles. This method is commonly used for sterilizing liquids or separating substances of different molecular sizes.
6. **Crossflow filtration:** Crossflow filtration, also known as tangential flow filtration, involves the parallel flow of the mixture and the liquid passing through the filter surface. It is particularly useful for separating particles that tend to clog traditional filters, such as colloidal suspensions or viscous liquids.

1.3. Applications

☆ The ability to selectively separate particles and substances based on size, shape, and composition makes filtration an indispensable technique in various fields, providing solutions for purification, separation, and environmental protection.

☆ A few examples of the applications of filtration are listed below.

1. **Water treatment:** Filtration is widely used in water treatment processes to remove impurities, sediment, and microorganisms, ensuring safe and clean drinking water.
2. **Biotechnology and biopharmaceuticals:** Filtration is an essential step in the production of biotechnology and biopharmaceutical products, including cell culture, fermentation, protein purification, and virus removal.
3. **Air filtration:** Filtration is employed in HVAC systems, industrial processes, and cleanrooms to remove airborne particles, pollutants, and allergens from the air, ensuring better indoor air quality.
4. **Pharmaceutical industry:** Filtration is crucial in pharmaceutical manufacturing for the separation and purification of drug substances, removal of impurities, and sterilization of liquids, gases, and air.

5. **Food and beverage industry:** Filtration is used in the production of food and beverages to remove particles, clarify liquids, and sterilize ingredients, ensuring product safety and quality.
6. **Environmental monitoring:** Filtration techniques are employed in environmental monitoring to collect and analyze particulate matter, microorganisms, and pollutants in air, water, and soil samples.
7. **Oil and gas industry:** Filtration is utilized in the oil and gas industry for the separation of impurities, solids, and water from crude oil and natural gas streams, ensuring product quality and pipeline integrity.
8. **Laboratory research:** Filtration is commonly used in laboratories for sample preparation, clarification of solutions, removal of particulates, and sterilization of media and solutions.
9. **Mining and chemical industry:** Filtration is employed in the mining and chemical industries for the separation and recovery of valuable solids, removal of impurities, and wastewater treatment.
10. **Automotive and aerospace industries:** Filtration is utilized in these industries for the purification of fuels, oils, and hydraulic fluids, ensuring the proper functioning and longevity of engines and equipment.

2. Distillation

- ☆ Distillation is used for the separation and purification of substances based on differences in boiling points.
- ☆ It is commonly employed in various industries and laboratory settings for the separation and purification of substances.
- ☆ Distillation plays a crucial role in processes such as chemical synthesis, petroleum refining, and the production of alcoholic beverages.
- ☆ Distillation offers several advantages as a separation technique. It is a versatile and widely applicable method that allows for the separation of complex mixtures into their individual components. Distillation can achieve high levels of purity and is capable of producing large quantities of purified substances. It is also a non-destructive process, as the components retain their chemical properties throughout the separation.
- ☆ However, there are certain limitations to distillation. It is primarily applicable to liquid mixtures and may not be suitable for separating substances that decompose at high temperatures or have extremely close boiling points. Additionally, distillation can be energy-intensive, especially in cases where a large amount of heat is required for separation.

2.1. Principle

- ☆ The principle behind distillation is the conversion of a liquid mixture into vapour by heating and then condensing the vapour back into liquid form.
- ☆ The mixture is heated in a distillation apparatus, and as the temperature rises, the component with the lower boiling point vaporizes first, leaving behind the component with the higher boiling point. The vapour is then collected and cooled, causing it to condense back into liquid form. This condensed liquid, known as the distillate, is separated from the remaining liquid mixture.

2.2. Types

- ☆ There are several types of distillation techniques commonly used:

1. **Simple distillation:** Simple distillation is suitable for separating liquids with a large difference in boiling points. It involves a single distillation apparatus, and the vapour is directly condensed and collected. Simple distillation is commonly used for the separation of volatile compounds from non-volatile impurities.
2. **Fractional distillation:** Fractional distillation is employed when the boiling points of the components in the mixture are close together. It involves the use of a fractionating column, which provides a large surface area for vapour-liquid contact. The column helps to separate the components more effectively by allowing repeated vaporization and condensation. Fractional distillation is used for the separation of complex mixtures, such as petroleum fractions and alcoholic beverages.
3. **Vacuum distillation:** Vacuum distillation is employed when the boiling points of the components are high or when heat-sensitive compounds need to be separated. By reducing the pressure in the distillation apparatus, the boiling points of the components are lowered, minimizing the risk of thermal degradation. Vacuum distillation is commonly used in industries such as pharmaceuticals, where delicate compounds need to be isolated.
4. **Steam distillation:** Steam distillation is utilized for the separation of heat-sensitive compounds or those with high boiling points. It involves passing steam through the mixture, causing the volatile components to vaporize along with the steam. The vapour is then condensed and collected separately. Steam distillation is commonly used in the extraction of essential oils from plants.

2.3. Applications

- ☆ The versatility, effectiveness, and reliability of distillation make it a vital technique in various industries, enabling the separation, purification, and concentration of substances for different purposes.
- ☆ A few examples of the applications of distillation are listed below.

1. **Petroleum refining:** Distillation is an integral part of the petroleum refining process, where crude oil is separated into different fractions based on their boiling points. This allows for the production of various products such as gasoline, diesel, jet fuel, and lubricants.
2. **Alcohol production:** Distillation plays a key role in the production of alcoholic beverages. Fermented liquids are distilled to separate ethanol from other compounds, resulting in higher alcohol content and improved purity.
3. **Chemical synthesis:** Distillation is used in chemical synthesis to purify products and separate desired compounds from reaction mixtures. It enables the removal of impurities, by-products, or unreacted substances, resulting in higher product purity.
4. **Water purification:** Distillation is an effective method for purifying water by separating it from impurities, contaminants, and dissolved solids. This process is commonly used in desalination plants and in areas where access to clean water is limited.
5. **Essential oil extraction:** Steam distillation is widely employed in the extraction of essential oils from plants. The steam carries the volatile compounds from the plant material, which are then condensed and collected separately.
6. **Pharmaceutical industry:** Distillation is used in the pharmaceutical industry for the purification and isolation of active pharmaceutical ingredients (APIs) and other compounds. It allows for the removal of impurities and the separation of desired substances.
7. **Food industry:** Distillation is used in food processing to remove impurities, separate flavours and fragrances, and concentrate desired components. It is employed in the production of various food products, such as juices, extracts, and flavours.
8. **Environmental applications:** Distillation plays a role in environmental applications, such as the treatment of industrial wastewater and the recovery of solvents for reuse. It allows for the separation and recovery of valuable compounds while reducing environmental pollution.

3. Extraction

- ☆ Extraction is widely used in various scientific fields and industries to separate and isolate desired components from complex mixtures.
- ☆ It involves the transfer of a target substance from one phase to another, typically from a solid or liquid phase into a liquid or gas phase.
- ☆ Extraction techniques are based on the differences in solubility and affinity of the components in the mixture for different solvents or phases.

3.1. Principle

- ☆ Extraction is based on the principles of solubility and affinity.
- ☆ It involves the transfer of a target substance from one phase to another, utilizing the differences in their solubility or affinity for different solvents or phases.
- ☆ The process of extraction relies on the fact that different substances have varying degrees of solubility in different solvents, allowing for the selective separation of desired components from a mixture.
- ☆ By choosing appropriate solvents and optimizing extraction conditions, it is possible to achieve efficient separation and isolation of specific substances.

3.2. Types

- ☆ Extraction methods can be categorized into liquid-liquid extraction, solid-phase extraction, and supercritical fluid extraction.
- ☆ Each method has its unique principles and applications, allowing for the efficient separation and purification of substances based on their physical and chemical properties.

1. **Liquid-liquid extraction:** Liquid-liquid extraction, also known as solvent extraction, involves the transfer of a solute from one liquid phase (the feed phase) into another liquid phase (the extractant phase). This method relies on the difference in solubility of the components in the mixture between the two immiscible liquid phases. Common solvents used in liquid-liquid extraction include organic solvents such as chloroform, ethyl acetate, and diethyl ether. Liquid-liquid extraction finds applications in various industries, including pharmaceuticals, chemicals, and environmental analysis.
2. **Solid-phase extraction:** Solid-phase extraction (SPE) is a technique that uses solid sorbents to selectively extract and concentrate target compounds from a liquid sample. The target compounds are adsorbed onto the solid sorbent based on their affinity and interactions with the sorbent material.

After extraction, the sorbent is eluted with a suitable solvent to obtain the desired compounds. Solid-phase extraction is commonly used in analytical laboratories for sample preparation and purification before analysis.

3. **Supercritical fluid extraction:** Supercritical fluid extraction (SFE) utilizes supercritical fluids, such as carbon dioxide (CO_2), as the extracting solvent. Supercritical fluids exhibit unique properties, combining the characteristics of both gases and liquids. Under specific temperature and pressure conditions, the supercritical fluid can effectively dissolve and extract target compounds. Supercritical fluid extraction is known for its environmentally friendly nature, as supercritical CO2 is non-toxic and easily recoverable. It is widely used in the extraction of natural products, food processing, and pharmaceutical manufacturing.

3.3. Applications

☆ Extraction has numerous applications, some of which are outlined below.

1. **Chemical industry:** Extraction plays a crucial role in the chemical industry for the separation and purification of desired compounds from complex mixtures. It enables the extraction of specific chemicals, such as organic acids, alkaloids, and natural products, from raw materials.
2. **Pharmaceutical industry:** Extraction is extensively used in the pharmaceutical industry for the isolation and purification of active pharmaceutical ingredients (APIs) from natural sources or synthesis mixtures. It allows for the separation of target compounds, removal of impurities, and concentration of APIs for drug formulation.
3. **Biotechnology:** Extraction is crucial in biotechnology for the isolation and purification of biomolecules, such as proteins, enzymes, and nucleic acids, from cells and tissues. It enables the production of high-quality biomolecules for research, diagnostics, and therapeutic applications.
4. **Environmental analysis:** Extraction techniques are employed in environmental analysis to extract pollutants, contaminants, and organic compounds from water, soil, and air samples. This enables the identification and quantification of pollutants, contributing to environmental monitoring and assessment.
5. **Food and beverage industry:** Extraction is utilized in the food and beverage industry for the extraction of flavours, colours, and bioactive compounds from natural sources. It allows for the production of extracts, essential oils, and concentrated ingredients used in the formulation of food products.

6. **Forensic science:** Extraction techniques are employed in forensic science for the extraction and isolation of drugs, toxins, and other compounds from biological samples. This aids in the analysis and identification of substances in criminal investigations and toxicology studies.
7. **Environmental remediation:** Extraction techniques are utilized in environmental remediation to extract and remove contaminants from soil and water. This process helps in the purification and restoration of polluted sites.
8. **Petrochemical industry:** Extraction is employed in the petrochemical industry for the separation of different components from crude oil and natural gas. It enables the extraction of valuable hydrocarbons and the removal of impurities, contributing to the production of fuels and chemicals.
9. **Cosmetics and fragrance industry:** Extraction techniques are used in the cosmetics and fragrance industry to extract essential oils, aromas, and natural compounds from plants and botanical sources. These extracts are then incorporated into cosmetic formulations and fragrances.
10. **Mining industry:** Extraction techniques are employed in the mining industry for the separation and concentration of valuable minerals from ores. This allows for the extraction of metals and minerals used in various industrial processes.

4. Evaporation

- ☆ Evaporation involves the conversion of a liquid into a gaseous state by heating or reducing the pressure.
- ☆ It is based on the principle that different substances have different boiling points, and when a mixture is heated, the component with the lowest boiling point evaporates first, leaving behind the higher boiling point components.
- ☆ The process of evaporation is widely employed in various industries and laboratory settings for purposes such as concentration, purification, and isolation of substances.
- ☆ By harnessing the selective vaporization of components, evaporation allows for the separation of desired substances from complex mixtures.

4.1. Principle

- ☆ The principle of evaporation is based on the concept that different substances have different boiling points.

☆ When a mixture is heated or subjected to reduced pressure, the component with the lowest boiling point evaporates first, while the higher boiling point components remain in the liquid phase. This selective vaporization allows for the separation of desired substances from a mixture.

☆ Evaporation occurs when the energy supplied to the liquid overcomes the intermolecular forces holding the molecules together, enabling them to escape into the gas phase. The process involves the conversion of a liquid into a vapour, which can then be collected and condensed to obtain a purified or concentrated substance.

4.2. Types

☆ Depending on the desired outcome and the nature of the mixture, different types of evaporation methods can be employed.

☆ The choice of evaporation method depends on factors such as the properties of the mixture, desired separation goals, energy efficiency, and the specific requirements of the application.

☆ By selecting the appropriate type of evaporation, it is possible to achieve efficient separation, concentration, and purification of substances.

☆ Some common types of evaporation techniques include:

1. **Simple evaporation:** This is the most basic form of evaporation, where a liquid is heated to its boiling point to allow for the evaporation of the solvent, leaving behind the solute or desired substance.

2. **Forced circulation evaporation:** In this method, a mechanical circulator is used to enhance the transfer of heat and improve the efficiency of the evaporation process. It is particularly useful for handling high-viscosity or heat-sensitive solutions.

3. **Multiple-effect evaporation:** This technique involves utilizing a series of evaporators, where the vapour generated in one evaporator is used as the heating source for the next stage. This results in increased energy efficiency and higher concentration of the desired substance.

4. **Vacuum evaporation:** By applying reduced pressure, vacuum evaporation lowers the boiling point of the liquid, allowing for evaporation at lower temperatures. It is useful for heat-sensitive materials and enables the concentration of volatile components.

5. **Flash evaporation:** Also known as flash distillation, this method involves rapidly heating a liquid mixture and then suddenly reducing the pressure to induce instantaneous vaporization. It is used for the separation of volatile components from non-volatile ones.

6. **Thin film evaporation:** In this technique, a thin film of liquid is spread over a heated surface, maximizing the surface area for evaporation and increasing the efficiency of the process. Thin film evaporators are commonly used for the concentration of heat-sensitive substances.

4.3. Applications

☆ Evaporation has numerous applications in various industries and scientific fields.

☆ It is employed for concentration, purification, and separation purposes, offering several advantages in terms of efficiency, versatility, and cost-effectiveness.

☆ Some key applications of evaporation are:

1. **Water treatment:** Evaporation is extensively used in wastewater treatment processes. By evaporating water from wastewater streams, impurities and contaminants are left behind, resulting in concentrated waste that can be further treated or disposed of safely.
2. **Food and beverage industry:** Evaporation is employed for the concentration and preservation of food and beverage products. It is used to remove excess water, thereby increasing the concentration of flavours, sugars, and other desired components in the final product.
3. **Pharmaceutical industry:** Evaporation plays a crucial role in the production of drugs and pharmaceuticals. It is utilized for solvent removal, concentration, and purification of drug compounds, enabling the isolation of pure active pharmaceutical ingredients (APIs).
4. **Chemical industry:** Evaporation is widely used in chemical processes for solvent recovery and concentration of chemical solutions. It is employed in the production of various chemicals, including polymers, solvents, and speciality chemicals.
5. **Environmental sciences:** Evaporation techniques are employed for the treatment of industrial effluents, particularly in industries where wastewater management is crucial. By evaporating water, pollutants and contaminants can be separated and concentrated for further treatment or disposal.
6. **Energy production:** Evaporation plays a significant role in energy production processes such as desalination and the generation of solar power. It is used in the production of freshwater from seawater by evaporating the water and condensing it as pure freshwater. In solar power plants, evaporation is employed for cooling purposes.

7. **Materials research:** Evaporation is utilized in the fabrication of thin films and coatings in materials science and research. By evaporating a solution or material onto a substrate, thin films with specific properties and thicknesses can be deposited.

5. Crystallization

- ☆ Crystallization is a widely employed technique for the purification and separation of solids from liquids.
- ☆ It is based on the principle of the formation of ordered solid crystals from a supersaturated solution or melt.
- ☆ Crystallization is a versatile technique that finds applications in various industries, including pharmaceuticals, chemicals, food, and materials science.

5.1. Principle

- ☆ The principle of crystallization is based on the fact that solutes can form ordered and well-defined crystals from a supersaturated solution or melt.
- ☆ The process involves the controlled reduction of solubility by manipulating temperature, pressure, or solvent composition.
- ☆ When the solute concentration exceeds its saturation point, nucleation occurs, leading to the formation of tiny crystal nuclei. These nuclei then grow by attracting additional solute molecules from the solution, resulting in larger and well-formed crystals.
- ☆ The principle of crystallization relies on the differences in solubility between the desired solute and impurities, allowing for the selective separation and purification of substances.
- ☆ By controlling the crystallization conditions, such as the cooling rate or seeding, it is possible to obtain crystals with desired properties and high purity.
- ☆ The process of crystallization involves the following steps:

1. **Solution preparation:** A solute is dissolved in a suitable solvent to form a solution. The solute can be a pure compound or a mixture of different substances.
2. **Supersaturation:** The solution is heated or cooled to reach a state of supersaturation, where the concentration of the solute exceeds its solubility limit at a given temperature.
3. **Nucleation:** Under controlled conditions, the supersaturated solution is induced to form small crystalline particles called nuclei. Nucleation can occur spontaneously or be initiated by adding a seed crystal or a foreign particle.

4. **Crystal growth:** The nuclei act as sites for further crystal growth. As the solute molecules attach to the existing nuclei, the crystals grow in size and become more defined.
5. **Separation:** Once the crystals have reached a desirable size, they are separated from the remaining solution through filtration, centrifugation, or other solid-liquid separation methods.

5.2. Types

☆ Crystallization can be classified into several types based on the specific process and conditions employed. The two main types of crystallization are:

1. **Cooling crystallization:** This is the most commonly used type of crystallization. It involves the gradual cooling of a solution or melts, causing the solute to become less soluble and form crystals. As the temperature decreases, the solute molecules start to aggregate and organize into ordered crystal structures. The crystals can be further purified and separated through filtration or other solid-liquid separation methods.
2. **Evaporative crystallization:** In this type of crystallization, the solute is precipitated by evaporating the solvent from the solution. As the solvent evaporates, the solute concentration increases, reaching a supersaturated state and promoting crystal formation. The crystals can be collected by filtration or centrifugation.

☆ Other types of crystallization techniques include:

3. **Fractional crystallization:** This method involves the sequential crystallization of different compounds from a solution. By carefully controlling the temperature and solubility conditions, multiple fractions of different compounds can be obtained.
4. **Solvent crystallization:** Solvent crystallization involves the addition of a second solvent that is miscible with the primary solvent but has a lower solubility for the solute. The solute precipitates out of the solution as crystals due to the reduced solubility in the mixed solvent system.
5. **Slow cooling crystallization:** This technique focuses on controlling the cooling rate to encourage the formation of large and well-defined crystals. By cooling the solution at a slow and controlled rate, the solute has more time to organize into regular crystal structures.
6. **Freeze crystallization:** Freeze crystallization utilizes freezing to separate the solute from the solvent. By subjecting the solution to freezing temperatures, ice crystals form, and the solute concentrates in the remaining liquid. The ice crystals can be removed, leaving a more concentrated solute solution.

- ☆ These various types of crystallization techniques offer flexibility in terms of process design and optimization.
- ☆ The choice of the appropriate method depends on factors such as the solute properties, desired crystal characteristics, and overall process requirements.

5.3. Applications

- ☆ Crystallization has diverse applications in various industries.
- ☆ It finds use in research laboratories, academic institutions, and various other fields where the separation and purification of substances are essential.
- ☆ Some key applications of crystallization include:

1. **Pharmaceutical industry:** Crystallization plays a crucial role in the production of pharmaceutical compounds. It is used for the purification and isolation of active pharmaceutical ingredients. By controlling the crystallization process, pharmaceutical companies can obtain high-purity crystals suitable for drug formulation.
2. **Food and beverage industry:** Crystallization is employed in the food industry for the production of sugar, salt, and other crystalline food products. It is used to obtain refined sugar crystals, salt crystals, and other desired forms of food ingredients. Crystallization helps improve the texture, appearance, and shelf life of food products.
3. **Chemical industry:** Crystallization is widely utilized for the separation and purification of chemical compounds. It is employed to remove impurities, separate isomers, and isolate specific chemical substances. Crystallization is particularly useful in industries such as petrochemicals, fine chemicals, and speciality chemicals.
4. **Materials science:** Crystallization is crucial in the field of materials science for the production of high-quality crystals with desired properties. It is used to synthesize single crystals for electronic devices, semiconductors, optical components, and other advanced materials. Crystallization techniques help control crystal structure, size, and purity, which are critical for achieving the desired material properties.
5. **Environmental applications:** Crystallization is also employed in environmental applications. It is used for wastewater treatment to remove impurities and recover valuable substances. Crystallization can effectively separate and recover metals, salts, and other compounds from industrial effluents.

6. **Mining and mineral processing:** Crystallization techniques are utilized in mining and mineral processing industries to extract and purify valuable minerals. It is used to separate minerals from ore and obtain high-purity crystal forms Crystallization helps improve the efficiency and effectiveness of mineral extraction processes.

6. Precipitation

- ☆ Precipitation involves the formation of a solid precipitate from a solution by introducing a reactant or changing the environmental conditions.
- ☆ This technique is based on the principle of the solubility of substances in a solvent.
- ☆ By manipulating the solubility, the desired compound can be separated and recovered in solid form.
- ☆ Precipitation finds wide applications in various industries, research laboratories, and academic settings.
- ☆ The process of precipitation involves several steps. First, a reactant is added to the solution, which reacts with the dissolved compound to form an insoluble precipitate. The reactant can be a chemical compound that reacts with the dissolved substance to form an insoluble compound or a change in pH, temperature, or concentration that causes the compound to become insoluble. The precipitate is then allowed to settle, and the supernatant liquid is separated from the solid using techniques such as filtration, centrifugation, or decantation. The precipitate can be further washed and dried to obtain the pure solid compound.

6.1. Principle

- ☆ The principle of precipitation is based on the solubility of substances in a given solvent or solution. When certain conditions are met, such as the addition of a reactant or changes in temperature, pH, or concentration, the solubility of a compound can be altered, leading to the formation of a solid precipitate.
- ☆ The principle of precipitation relies on the concept of chemical reactions. By introducing a reactant that forms an insoluble compound with the target substance, a precipitate can be generated. This can be achieved by exploiting the differences in chemical properties between the compound of interest and the reactant. For example, in the case of metal ions, precipitation can occur by adding a precipitating agent that forms an insoluble metal salt.

- ☆ The solubility of compounds is influenced by factors such as temperature, pressure, concentration, and pH. By manipulating these variables, the conditions can be optimized to induce precipitation. Lowering the solubility of a compound can be achieved by decreasing the temperature or increasing the concentration of another compound that interacts with it to form an insoluble product.
- ☆ Precipitation reactions are often guided by the principles of thermodynamics and chemical equilibrium. The formation of a precipitate occurs when the concentration of the reactants exceeds the solubility product, which is the equilibrium constant for the dissolution and precipitation reaction. The precipitation process reaches a state of equilibrium where the rate of precipitation is equal to the rate of dissolution.
- ☆ Control over the precipitation process is crucial to obtain the desired purity and yield of the precipitate. Factors such as the rate of mixing, reaction time, pH adjustment, and temperature control play significant roles in achieving efficient precipitation.

6.2. Types

- ☆ There are several types of precipitation techniques that are used in chemical separation processes. These techniques involve different mechanisms and strategies to induce the formation of a solid precipitate.
- ☆ Each type of precipitation technique has its own advantages and limitations, and the choice of method depends on the specific requirements of the separation process and the properties of the compounds involved.
- ☆ The selection of the appropriate precipitation method is crucial to achieving the desired separation efficiency and yield of the precipitate.
- ☆ Some common types of precipitation methods include:

1. **Chemical precipitation:** In this method, a chemical reactant is added to a solution to form an insoluble compound with the target substance. The reactant interacts with the compound of interest, leading to the formation of a solid precipitate.
2. **Double decomposition:** This precipitation method involves the reaction between two soluble salts to produce a precipitate of an insoluble compound. It is based on the principle of metathesis, where ions exchange between the two reactants to form a new compound that is less soluble and precipitates out.

3. **pH-Induced precipitation:** Changing the pH of a solution can alter the solubility of certain compounds. By adjusting the pH to a specific range, the compound of interest can become insoluble and precipitate out of the solution.

4. **Temperature-induced precipitation:** Temperature plays a significant role in the solubility of substances. Lowering the temperature of a solution can reduce the solubility of certain compounds, causing them to precipitate.

5. **Solvent precipitation:** This method involves adding a solvent to the solution to induce the precipitation of the compound of interest. The addition of the solvent changes the solvent power of the solution, leading to the formation of a solid precipitate.

6.3. Applications

☆ Precipitation is employed in various industries and research fields for different purposes.

☆ It enables the separation and purification of compounds, the removal of contaminants from solutions, and the synthesis of specific materials.

☆ The precise control over reaction conditions and choice of reactants allows for the customization of precipitation processes to meet specific requirements in various applications.

☆ Some key applications of precipitation include:

1. **Research and laboratory applications:** Precipitation is commonly used in research laboratories and academic settings for a range of applications. It is employed for the separation and purification of compounds, synthesis of nanoparticles, protein precipitation, and DNA/RNA isolation.

2. **Pharmaceutical industry:** Precipitation is used for the purification and isolation of pharmaceutical compounds. It allows the separation of desired drug substances from impurities or unwanted by-products. Precipitation techniques such as salting out and co-precipitation are utilized in pharmaceutical manufacturing processes.

3. **Water and wastewater treatment:** Precipitation is an essential step in water and wastewater treatment processes. It is used to remove heavy metals, suspended particles, and other contaminants from water sources. Precipitation techniques help to precipitate the contaminants, which can then be removed through filtration or sedimentation.

4. **Mining and metallurgy:** Precipitation is utilized in the mining and metallurgical industry for the extraction and purification of metals. It is employed to separate desired metals from their ores or solutions. Precipitation techniques are used to obtain high-purity metal compounds that can be further processed into metals.

5. **Chemical synthesis:** Precipitation plays a crucial role in chemical synthesis for the formation of specific compounds. It is employed to produce solid substances with desired properties, such as particle size, morphology, and purity. Precipitation reactions are widely used in the synthesis of catalysts, pigments, and various speciality chemicals.

6. **Environmental analysis:** Precipitation techniques are utilized in environmental analysis for the determination of trace elements and pollutants in samples. The precipitated compounds can be collected and analyzed using various analytical methods to assess environmental quality and monitor contamination levels.

7. Electrolysis

- ☆ Electrolysis involves the use of an electric current to drive a non-spontaneous chemical reaction.
- ☆ It involves an electrolytic cell comprising an anode and a cathode immersed in an electrolyte solution or molten salt. The flow of electric current triggers oxidation at the anode and reduction at the cathode, leading to the separation of substances. Electrolysis is commonly employed in metal extraction, electroplating, and the production of chemicals.

7.1 Principle

- ☆ The principle of electrolysis is based on the concept of electrochemical reactions.
- ☆ It involves the use of an electric current to drive non-spontaneous chemical reactions.
- ☆ The process occurs in an electrolytic cell where an anode and a cathode are immersed in an electrolyte solution or molten salt. When a direct current is applied, oxidation occurs at the anode, resulting in the release of electrons and the generation of positive ions. Simultaneously, reduction occurs at the cathode, where positive ions are attracted and gain electrons, forming neutral or negative species. This transfer of electrons and ions enables the separation and purification of substances during electrolysis.

7.2. Types

- ☆ Electrolysis, as a chemical separation technique, can be classified into two main types: galvanic (voltaic) electrolysis and electrolytic electrolysis.

1. **Galvanic (voltaic) electrolysis:** In this type of electrolysis, the chemical reaction is spontaneous, and electric energy is generated as a result of the reaction. Galvanic cells, also known as voltaic cells, convert chemical energy into electrical energy. These cells consist of two half-cells, each with

an electrode immersed in an electrolyte solution. The chemical reaction between the electrodes generates an electric current. Examples of galvanic electrolysis include batteries and fuel cells.

2. **Electrolytic electrolysis:** In this type of electrolysis, the chemical reaction is non-spontaneous, and an external electric current is applied to drive the reaction. The electrolytic cell contains an anode (positive electrode) and a cathode (negative electrode), which are connected to a power supply. When the electric current is passed through the cell, the anode undergoes oxidation, and the cathode undergoes reduction. Electrolytic electrolysis is commonly used for various applications such as metal extraction, electroplating, and the production of chemicals. Different types of electrolytic electrolysis can be classified based on the specific processes involved:

    a) **Industrial electrolysis:** This type of electrolysis is used for large-scale applications such as the production of chlorine, sodium hydroxide, and aluminium. Industrial electrolytic cells are designed to handle high currents and produce the desired products efficiently.

    b) **Electrorefining:** It is a process used to purify metals by electrolysis. Impure metals are used as anodes, and the desired metal is deposited on the cathode, resulting in the purification of the metal.

    c) **Electroplating:** It involves the deposition of a thin layer of metal onto a surface by electrolysis. This process is widely used for decorative and protective coatings on various materials.

    d) **Electrolysis for water splitting:** Electrolysis of water involves the decomposition of water molecules into hydrogen gas and oxygen gas. This process is essential for hydrogen production and energy storage.

7.3 Applications

☆ Electrolysis has numerous applications in various industries. Here are some of the key applications of electrolysis:

1. **Electroplating:** Electrolysis is extensively used in electroplating processes to apply a thin layer of metal onto a substrate. This technique is widely employed in industries such as automotive, electronics, and jewellery manufacturing to enhance the appearance, provide corrosion resistance, and improve the conductivity of surfaces.

2. **Metal extraction:** Electrolysis plays a crucial role in the extraction of metals from their ores. For example, electrolysis is employed in the extraction of aluminium from bauxite ore, copper from copper sulphide ores, and the refining of impure metals like gold and silver.

3. **Water electrolysis:** Electrolysis of water is used to produce hydrogen and oxygen gases. Hydrogen gas is a clean and versatile fuel that can be used in fuel cells for power generation, transportation, and industrial applications. The oxygen gas produced can be used in various industrial processes, including medical applications and wastewater treatment.

4. **Chlor-alkali industry:** Electrolysis is extensively used in the chlor-alkali industry for the production of chlorine gas, sodium hydroxide (caustic soda), and hydrogen gas. These chemicals are essential for manufacturing a wide range of products, such as bleach, PVC, detergents, and paper.

5. **Electrolytic refining:** Electrolysis is employed for the purification and refining of metals. Impure metals are used as anodes, and through electrolysis, the desired metal is deposited on the cathode, resulting in the purification of the metal. This technique is commonly used to refine copper, zinc, and other metals.

6. **Electrolysis in electrochemical cells:** Electrolysis is central to the functioning of various electrochemical cells, including batteries and fuel cells. In these devices, chemical reactions involving the transfer of electrons and ions occur, converting chemical energy into electrical energy for power storage and generation.

7. **Waste treatment:** Electrolysis is utilized in wastewater treatment processes, particularly in the removal of heavy metals and organic compounds. Electrochemical methods are employed to break down and neutralize pollutants, contributing to the purification of water resources.

8. Solvent extraction

- ☆ Solvent extraction involves the selective transfer of a target compound from one phase to another using a solvent.
- ☆ It is also known as liquid-liquid extraction.
- ☆ This technique is based on the differences in solubility and distribution of compounds between immiscible phases.
- ☆ It finds extensive applications in various industries and laboratories for the extraction, purification, and separation of a wide range of organic and inorganic compounds.
- ☆ The principle behind solvent extraction is that different compounds have varying affinities for solvents based on their chemical properties, such as polarity and solubility. By choosing an appropriate solvent and adjusting the extraction conditions, it is possible to selectively extract a target compound while leaving unwanted impurities behind.

8.1 Principle

- ☆ The principle of solvent extraction, also known as liquid-liquid extraction, is based on the differential solubility of compounds in two immiscible liquid phases.
- ☆ The technique relies on the fact that different compounds exhibit different affinities for the solvent phases, allowing for the selective separation and extraction of target compounds from a mixture.
- ☆ The principle of solvent extraction can be understood through the concept of distribution or partitioning. When a mixture is brought into contact with two immiscible solvents, the compounds in the mixture will distribute themselves between the two phases based on their solubility properties. The target compound of interest will preferentially dissolve in one of the solvents, while the impurities or unwanted compounds will remain in the other solvent.
- ☆ The selectivity of solvent extraction is determined by several factors, including the choice of solvents, the pH of the solution, the temperature, and the extraction time. By manipulating these parameters, it is possible to optimize the separation and extraction process and achieve high levels of selectivity and efficiency.

8.2. Types

- ☆ Solvent extraction encompasses several types based on the specific characteristics and requirements of the separation process.
- ☆ Each type offers its own advantages and is chosen based on the specific requirements of the separation process, including the nature of the target compound, the complexity of the mixture, and the desired level of separation and purity.
- ☆ Some common types of solvent extraction include:

1. **Simple solvent extraction:** This is the most basic type of solvent extraction, where a solute is transferred from one solvent phase to another. It involves the immiscibility of two solvents, typically an organic solvent and an aqueous phase. The solute is selectively partitioned between the two phases based on its solubility properties.
2. **Countercurrent extraction:** This technique involves multiple stages or steps of solvent extraction, where the extracted solute is continuously transferred from one solvent phase to another. The goal is to achieve a higher degree of separation and concentration of the target compound by maximizing the contact between the two phases.

3. **Solid-liquid extraction:** In this type of solvent extraction, the target compound is extracted from a solid matrix into a liquid solvent. This technique is commonly used for the extraction of natural products, such as essential oils, from plant materials.
4. **Supercritical fluid extraction:** Supercritical fluid extraction involves the use of a supercritical fluid, such as carbon dioxide, as the extracting solvent. Supercritical fluids have unique properties that combine the advantages of both liquids and gases, making them efficient solvents for certain applications, such as the extraction of heat-sensitive compounds.
5. **Microwave-assisted extraction:** This technique combines solvent extraction with microwave irradiation to enhance extraction efficiency. The use of microwave energy can accelerate the extraction process by promoting the penetration of the solvent into the sample matrix and facilitating the release of the target compounds.

8.3. Applications

- ☆ Solvent extraction finds numerous applications across various industries.
- ☆ Its ability to selectively extract and separate specific compounds from complex mixtures makes it a valuable tool in many processes.
- ☆ Some common applications of solvent extraction are:

1. **Pharmaceutical industry:** Solvent extraction is widely used in the pharmaceutical industry for the purification and isolation of active pharmaceutical ingredients from natural sources or reaction mixtures. It allows for the separation of desired compounds from impurities or unwanted by-products, ensuring the production of high-quality drugs.
2. **Environmental analysis:** Solvent extraction plays a crucial role in environmental analysis, particularly in the extraction and concentration of pollutants and contaminants from soil, water, and air samples. It enables the removal of target analytes from complex matrices for accurate analysis and monitoring of environmental pollution.
3. **Mining industry:** Solvent extraction is commonly employed in the mining industry for the extraction of metals from ores. It facilitates the separation and recovery of valuable metals, such as copper, nickel, and uranium, from the ore matrix. Solvent extraction is also utilized in the purification of metals during the refining process.
4. **Food and beverage industry:** Solvent extraction is used for the extraction of flavours, fragrances, and essential oils from natural sources, such as plants, herbs, and fruits. It helps in obtaining concentrated extracts for use in food products, beverages, and perfumes.

5. **Petrochemical industry:** Solvent extraction is employed in the petrochemical industry for the purification and separation of various components from crude oil and petroleum products. It aids in the removal of impurities, sulphur compounds, and other contaminants to obtain refined products.
6. **Nuclear industry:** Solvent extraction is utilized in the nuclear industry for the separation and purification of radioactive elements, such as uranium and plutonium. It is a critical process in nuclear fuel reprocessing and the production of radioisotopes for medical and industrial applications.
7. **Waste management:** Solvent extraction techniques are employed in waste management processes to recover valuable compounds from industrial waste streams. It allows for the separation and extraction of useful materials, reducing waste and promoting recycling and resource recovery.

Glossary of Common Terms

Chromatography

Adsorption chromatography: A type of chromatography where the stationary phase is a solid material that adsorbs the components of a mixture based on their interactions with the surface.

Affinity chromatography: A type of chromatography that uses a stationary phase with specific binding sites to separate and purify compounds based on their biological activity or affinity.

Analytical chromatography: A type of chromatography that is used to identify and quantify the components of a mixture by separating them based on their interaction with a stationary phase and a mobile phase.

Baseline: The level of detector response in the absence of a sample in a chromatographic system.

Calibration curve: A graph of detector response versus analyte concentration used to quantify analytes in a sample.

Chiral chromatography: A type of chromatography that separates enantiomers, or stereoisomers, using a chiral stationary phase.

Chromatogram: A graphical representation of the separated components of a mixture obtained by chromatography.

Chromatographic peak shape: The appearance of a peak in a chromatogram, which can provide information about the quality of the separation.

Column chromatography: A type of chromatography that uses a column packed with a stationary phase to separate and purify compounds based on their physical and chemical properties.

Column efficiency: A measure of the ability of a chromatographic column to separate components of a mixture.

Column volume: It refers to the total volume of the chromatography column that is occupied by the stationary phase

Countercurrent chromatography (CCC): A type of liquid-liquid chromatography that uses a liquid stationary phase and a liquid mobile phase that flow in opposite directions, allowing for high resolution and efficient separation of compounds.

Crystallization: A separation technique where a solid solute is precipitated from a solvent by cooling or evaporation, forming well-defined crystals that can be separated from the solution.

Cutoff size: It refers to the maximum size of molecules that can effectively pass through the pores or interact with the stationary phase of the chromatographic material.

Dead volume: The volume of a chromatographic system that is not used for separation.

Detection limit: The lowest concentration of an analyte that can be detected in a chromatographic system.

Detector: A device used to detect and measure the separated components of a sample in a chromatographic system.

Distillation: A separation technique that utilizes differences in boiling points to separate components of a liquid mixture by vaporizing and condensing them.

Dye-ligand chromatography: A type of chromatography that uses a dye molecule with a specific affinity for the target molecule as the stationary phase, allowing for high selectivity and efficiency in protein and enzyme purification.

Electrochromatography (EC): A hybrid technique that combines the principles of chromatography and electrophoresis, using an electric field to drive the movement of charged compounds through a stationary phase.

Electrolysis: A separation technique that uses an electric current to drive a non-spontaneous chemical reaction, resulting in the separation of ions or the deposition of a substance onto an electrode.

Eluent: The solvent or mobile phase moves a sample through a chromatographic column.

Elution: The process of moving a sample through a chromatographic column.

Evaporation: A separation technique that involves the vaporization of a solvent from a solution or mixture, leaving behind the dissolved solute or solid residue.

Extraction: A separation technique that involves the selective dissolution of a substance from a mixture into a solvent based on differing solubilities.

Fast protein liquid chromatography (FPLC): A specialized variant of the HPLC technique used for efficient separation, purification, and analysis of proteins.

Filtration: A separation technique that involves passing a mixture through a porous medium to separate solid particles from the liquid or gas phase.

Flash chromatography: A fast and efficient method for separating and purifying compounds from a mixture using a column filled with silica gel or other adsorbents.

Gas chromatography: A type of chromatography that uses a gas as the mobile phase and a solid or liquid stationary phase to separate volatile compounds.

Gel filtration chromatography (GFC): A type of chromatography used to separate and purify macromolecules based on their size and shape. It is also called size exclusion chromatography.

Gradient elution: A method of elution in which the composition of the mobile phase changes over time.

High-performance liquid chromatography (HPLC): A chromatographic technique that uses high pressure to move the mobile phase through the stationary phase for improved separation and efficiency.

Hydrophilic interaction chromatography (HILIC): A type of chromatography that separates polar compounds using a hydrophilic stationary phase and a polar mobile phase.

Hydrophobic interaction chromatography (HIC): A type of liquid chromatography that separates molecules based on their hydrophobicity by using a stationary phase with a hydrophobic surface and a mobile phase with varying salt concentrations.

Immunoaffinity chromatography (IAC): A type of chromatography that uses antibodies to selectively capture and purify a target molecule from a complex mixture. It is commonly used in protein and peptide purification.

Injection: The process of introducing a sample into a chromatographic system.

Internal standard: A known quantity of a substance added to a sample to control for variations in sample preparation and instrument response.

Isocratic elution: A method of elution in which the composition of the mobile phase is constant throughout the separation.

Liquid chromatography (LC): A type of chromatography that separates compounds based on their interaction with a liquid stationary phase and a mobile phase, typically a solvent.

Liquid chromatography-mass spectrometry (LC-MS): A technique that combines liquid chromatography with mass spectrometry to identify and quantify compounds in a sample.

Liquid-liquid chromatography (LLC): A type of chromatography that separates compounds based on their solubility in two immiscible liquids, typically water and an organic solvent. It is useful for separating polar and nonpolar compounds.

Mobile phase: The mixture of solvents used to elute a sample in a chromatographic system.

Multi-dimensional chromatography: A technique that combines two or more chromatographic separations, such as gas and liquid chromatography, to achieve higher resolution and more comprehensive analysis of complex mixtures.

Paper chromatography: A type of chromatography that uses a paper strip as the stationary phase and a liquid solvent as the mobile phase to separate and analyze mixtures of small molecules.

Partition chromatography: A type of chromatography where the stationary phase is a liquid that partitions the components of a mixture between two immiscible liquids.

Peak: A graphical representation of a separated component in a chromatogram.

Precipitation: A separation technique that involves the formation of a solid precipitate from a chemical reaction between two or more substances, resulting in their separation from the solution.

Preparative chromatography: A type of chromatography that is used to purify larger quantities of a target compound for downstream applications, such as drug discovery, by separating it from a complex mixture.

Pseudoaffinity chromatography: A type of affinity chromatography that uses a ligand with low specificity for the target molecule to increase its capture and purification efficiency. It is often used in protein purification.

Pyrolysis gas chromatography (Py-GC): A type of chromatography that involves heating a sample to high temperatures to break it down into smaller molecules, which are then separated and analyzed using gas chromatography.

Quantification limit: The lowest concentration of an analyte that can be reliably quantified in a chromatographic system.

Resolution: A measure of the degree of separation between two closely eluting components in a chromatographic system.

Retention factor (Rf): The ratio of the distance travelled by a compound from the starting line to the distance travelled by the solvent from the starting line in a chromatogram.

Retention time: The time a component takes to travel through a chromatographic column and is detected.

Reverse Phase chromatography (RPC): A type of liquid chromatography that separates molecules based on their hydrophobicity using a stationary phase with a hydrophobic surface and a polar mobile phase.

Reversed-phase chromatography: A type of chromatography in which the stationary phase is nonpolar, and the mobile phase is polar, used to separate polar compounds.

Sample matrix: The components in a sample other than the analytes of interest.

Sample preparation: The process of preparing a sample for chromatographic analysis.

Sample throughput: The number of samples that can be analyzed in a given amount of time in a chromatographic system.

Selectivity: The ability of a chromatographic system to separate different components in a sample.

Separation factor: A measure of the ability of a chromatographic system to separate two components.

Size-exclusion chromatography (SEC): A type of chromatography that separates compounds based on size and molecular weight using a porous stationary phase.

Solvent extraction: A separation technique that involves the transfer of a solute from one solvent phase to another based on the differences in solubility or partitioning behaviour between the phases.

Sorbent: The solid material used as the stationary phase in a chromatographic column.

Sorption: The process by which components in a sample are adsorbed onto the stationary phase.

Specificity: The ability of a chromatographic system to identify and quantify a specific analyte in a sample.

Stationary phase: The solid or liquid material used to separate the components of a sample in a chromatographic system.

Supercritical fluid chromatography (SFC): A type of chromatography that uses a supercritical fluid, typically carbon dioxide, as the mobile phase for separating and analyzing compounds. It is useful for analyzing thermally labile and non-volatile compounds.

Thin-layer chromatography (TLC): A type of chromatography that uses a thin layer of stationary phase on a solid support to separate components in a mixture.

Void volume: The volume of a chromatographic system that is not occupied by the stationary phase.

Centrifugation

Analytical centrifugation: The application of centrifugation for studying the physical properties and interactions of particles and molecules.

Centrifugal acceleration: The acceleration experienced by a sample in a centrifuge, expressed in multiples of the Earth's gravity (g), typically measured in terms of RCF.

Centrifugal concentrator: A device used to concentrate dilute solutions or extract desired components by subjecting them to centrifugal force.

Centrifugal elutriation: A technique that separates particles based on their size by subjecting them to a counter-flowing fluid under centrifugal force.

Centrifugal field strength: The strength of the force field generated by the spinning rotor in a centrifuge.

Centrifugal field: The force field created by the spinning rotor in a centrifuge.

Centrifugal filtration: A method that separates solids from liquids by applying centrifugal force to force the liquid through a filter while retaining the solids.

Centrifugal force: The outward force experienced by objects in rotation, leading to the separation of components in a mixture.

Centrifugal fractionation: A technique that separates a sample into fractions based on their sedimentation characteristics.

Centrifugal sedimentation coefficient: A measure of the sedimentation rate of particles or molecules in a centrifugal field.

Centrifugal time: The duration for which a sample is subjected to centrifugal force.

Centrifugation: A process of separating components of a mixture based on their density using centrifugal force.

Centrifuge balance: The distribution of samples and containers within the rotor to maintain rotor balance during centrifugation.

Centrifuge brake: A safety feature in centrifuges that rapidly decelerates the rotor to ensure safe sample retrieval.

Centrifuge safety lid: A protective lid that prevents the rotor from being accessed during operation and ensures safe centrifugation.

Centrifuge tube: A cylindrical container used to hold samples during centrifugation.

Centrifuge: A laboratory instrument used for centrifugation, consisting of a rotor that spins at high speeds to generate centrifugal force.

Continuous flow centrifugation: A method where the sample is continuously introduced into the rotor while the separated components are continuously withdrawn.

Counter-current centrifugation: A method that uses a liquid medium flowing counter-currently to the centrifugal force to improve separation efficiency.

Density gradient centrifugation: A technique that uses a density gradient to separate particles of different densities in a sample.

Density gradient formation: The process of creating a continuous gradient of varying density within a centrifuge tube or container using solutions of different densities.

Density gradient: A continuous gradient of varying density within a centrifuge tube or container used for separating particles based on their buoyant density.

Differential centrifugation: A method that separates particles based on their size and density by subjecting the sample to multiple rounds of centrifugation at different speeds.

Differential sedimentation: The separation of particles based on their different settling velocities under centrifugal force.

Equilibrium density gradient centrifugation: A method where particles are separated based on their buoyant density in a density gradient created within the centrifuge tube.

Fixed-angle rotor: A type of centrifuge rotor where the centrifuge tubes are held at a fixed angle relative to the rotor axis.

Isolation: The process of separating a specific component or target substance from a mixture using centrifugation.

Isopycnic centrifugation: A technique that separates particles solely based on their buoyant density in a density gradient.

Microcentrifuge: A small-sized centrifuge used for separating small volumes of samples at high speeds.

Particle size analysis: The measurement and characterization of particle sizes in a sample using centrifugation techniques.

Pellet: The solid material that settles at the bottom of a centrifuge tube after centrifugation.

Preparative centrifugation: The use of centrifugation to obtain larger quantities of purified samples for further analysis or downstream applications.

Relative centrifugal force (RCF): The measure of the acceleration experienced by a sample in a centrifuge, expressed in multiples of the Earth's gravity (g).

Rotor capacity: The maximum number and volume of centrifuge tubes or containers that a rotor can hold.

Rotor: The rotating component of a centrifuge that holds the centrifuge tubes or containers.

Sediment: The solid material that settles at the bottom of a centrifuge tube or container after centrifugation.

Sedimentation coefficient: A measure of the rate at which particles or molecules sediment in a centrifugal field, expressed in Svedberg units (S).

Sedimentation: The process by which particles in a suspension settle down under the influence of gravity or centrifugal force.

Speed (rpm): The rotational speed of the centrifuge rotor, measured in revolutions per minute.

Supernatant: The liquid portion remaining above the pellet after centrifugation.

Svedberg unit: A non-metric unit used in ultracentrifugation to measure the sedimentation rate of particles and characterize their size and shape.

Swinging-bucket rotor: A type of centrifuge rotor where the centrifuge tubes swing outward during centrifugation.

Ultracentrifugation: A high-speed centrifugation technique used for separating and purifying subcellular components and macromolecules.

Index